Math Facts

Survival Guide to Basic Mathematics

Second Edition

THEODORE JOHN SZYMANSKI

TOMPKINS CORTLAND COMMUNITY COLLEGE

PWS Publishing Company

I(T)P An International Thomson Publishing Company

Boston • Albany • Bonn • Cincinnati • Detroit • London • Madrid • Melbourne • Mexico City
New York • Paris • San Francisco • Singapore • Tokyo • Toronto • Washington

PWS PUBLISHING COMPANY
20 Park Plaza, Boston, MA 02116-4324

I(T)P™ **International Thomson Publishing**
   The trademark ITP is used under license.

Printed and bound in the United States of America.
   7 8 9 10—03 02

ISBN 0-534-94734-4

Sponsoring Editor: Susan McCulley Gay	Marketing Manager: Marianne Rutter
Developmental Editor: Elizabeth R. Deck	Cover Design: Kathleen Wilson
Editorial Assistant: Hattie Schroeder	Manufacturing Coordinator: Ellen Glisker
Production Editor: Kathleen Wilson	Printer and Binder: Courier Westford, Inc.

Preface

MATH FACTS was written for use in developmental mathematics courses to fill a need expressed by students for "anchor points" in basic mathematics. The guide began as an index card project that provided minimal but essential information on single mathematical concepts. A sixteen-semester collection of these cards resulted in **MATH FACTS**.

NEW TOPICS:

Angles	Association
Calculator Hints	Commutation
Complex Fractions	Cubes and Cube Roots
Discussion of Pi	Geometric Shapes
Negative Numbers	Order of Magnitude
Precision in Numbers	References to Math in Nature
Surface Area	Triangles
Unit Analysis	Uses of the Multiplication Table

The alphabetical listing offers students entry points into basic concepts and math terminology. Each page is a simple treatment of a single concept. Reference is made at the bottom of each page to related ideas. The language chosen is that of operational definitions that have proven to be successful with developmental and nontraditional students, particularly those who do not favorlinear-processing styles. Cross-references help clarify differences between terms that may easily be confused.

Students find that **MATH FACTS** reduces anxiety because they feel comfortable with its compact size and clear presentation.

MATH FACTS was designed as either a stand-alone shelf reference or for use with a textbook or workbook in basic math courses. **MATH FACTS** is also used by students enrolled in other courses, such as introductory algebra, biology, or chemistry where a brief review of basic math concepts is helpful.

This is the second edition of **MATH FACTS**. Feedback from students and teachers indicated a need for additional topics such as geometry terms and the expansion of concepts.

Since its appearance in 1992, **MATH FACTS** has found many nontraditional applications. Continuing Education departments have used the book in industry for upgrading workers' math skills. Some businesses have employed the book as a pre-hiring requirement or as part of an on-the-job training program. Several businesses have purchased **MATH FACTS** to distribute them as gifts. It has been used in GED programs throughout the country.

A companion book entitled **ALGEBRA FACTS** is currently available and follows the same format.

TJS
Trumansburg, New York

T a b l e   o f   C o n t e n t s

ABSOLUTE VALUE   Symbol | |

Description: A way to indicate that the value of any quantity is positive, regardless of its sign.

Uses: Distances are always positive, but may come out negative in calculations, so one may "take the absolute value" of a distance.

Example: The absolute value of - 5 = 5

$$| - 5 | = 5$$

Rule: Evaluate the expression within the | | sign following the rules of the ORDER OF OPERATIONS and make that resulting quantity positive.

Example   Evaluate | 6 - 10 - 8 |

Step 1.   Evaluate the expression *inside* the bars | |.

CONTINUED NEXT PAGE   1

$$| \, 6 - 10 - 8 \, | \; = \; | -12 \, |$$

Step 2.   Express the absolute value of - 12 as + 12

$$| -12 \, | \; = \; 12$$

Absolute values are used in the physical sciences when expressing changes in measurements where the change is important and not the direction.

Additional examples:

$$| \, 14 - 2 \times 8 \, | \; = \; 2 \qquad \text{(be sure to observe the}$$
order of operations here)

$$-| \, 6 \, | \; = \; -6$$

$$-| -8 \, | \; = \; -8$$

Notice that the - sign *outside* the absolute value bars makes the final value negative.

ANGLES

An angle is the measure of opening between two line segments (called rays) that originate from a single point. Angles may be measured in *degrees*. Two lines that are perpendicular meet at 90° (right angles).

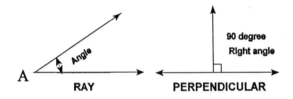

There are 360 degrees in a circle. A straight line is said to make an angle of 180° (degrees). This angle is often called a *straight* angle.

Angles measuring less than 90° are called *acute* angles. Angles greater than 90° are called *obtuse.*

CONTINUED NEXT PAGE

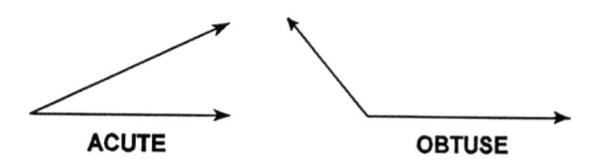

ACUTE   OBTUSE

When the sum of two angles equals 90° the angles are said to be *complementary.* One angle is the *complement* of the other.

When the sum of two angles equals 180° the angles are said to be *supplementary.* One angle is the *supplement* of the other.

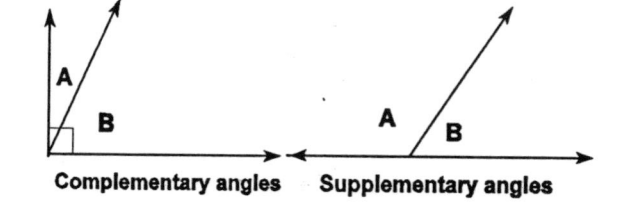

Complementary angles   Supplementary angles

4

AREA   Symbol: A

An area has two dimensions, *length* and *width*. Area is the measure of surface, flat objects such as carpets and walls. The units of area are *square* units such as *square* feet, *square* yards, *square* meters.

The area of almost any figure can be calculated by multiplying its length times its width.

FIGURE	FORMULA
SQUARE	side squared (length = width)
RECTANGLE	length x width
TRIANGLE	half length x width
CIRCLE	pi x square of radius
TRAPEZOID	average of widths x height

CONTINUED NEXT PAGE   5

Area is different from *PERIMETER*.　The peRIMeter is the distance *around* the RIM of an object. Perimeter is a LENGTH, a one dimensional measure. Perimeter is measured in inches, feet, meters, which are one-dimensional units.

Example:　Calculate the area and the perimeter of a garden plot that is 18' x 24'

$$\text{AREA} = \text{L} \times \text{W}$$

$$\text{AREA} = 18' \times 24' = 432 \text{ SQ FT}$$

$$\text{PERIMETER} = \text{L} + \text{L} + \text{W} + \text{W}$$

$$\text{PERIMETER} = 18' + 18' + 24' + 24' = 84'$$

AREA OF A CIRCLE

FORMULA: AREA = pi x R^2 or 3.14 x R^2 (R is the radius)

Example: Calculate the area of a circle that has a radius of 6"

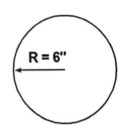

Step 1. Square the value of the radius.

$(6 \text{ in})^2$ = 36 square inches

Step 2. Multiply 36 square inches by 3.14

36 square inches x 3.14 = 113.04 square inches

Note that the units were squared along with the numbers. It is a good idea to carry along the units in your calculations.

If the diameter is given, it must be divided by 2 to get the radius.

CONTINUED NEXT PAGE

Example: Calculate the area of a circle that has a diameter of 17 cm.

Step 1. Divide the diameter by 2 to get the radius.

17cm divided by 2 is 8.5 cm.

Step 2. Square the 8.5 cm.

$(8.5 \text{ cm})^2 = 72.25 \text{ cm}^2$

Step 3. Multiply by 3.14

$72.25 \text{ cm}^2 \times 3.14 = 226.865 \text{ cm}^2$

Step 4. Round off to the *tenths* place.

226.9 square cm.

SEE ALSO: AREA, CIRCLES, TERMS

AREA OF A RECTANGLE

The area of a rectangle is given by L x W. Simply multiply the length by the width. Note that the units become square units. 3 feet times 7 feet equals 21 *square* feet. The units are important.

Example: Calculate the area of a rectangle that has a length of 7 feet and a width of 3 feet.

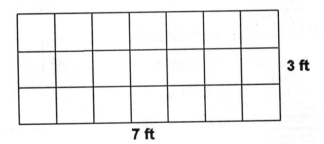

CONTINUED NEXT PAGE

Example: Calculate the area and the perimeter of a board that measures 12" x 17"

AREA = L x W

AREA = 12" x 17" = 204 sq in

PERIMETER = L + L + W + W

PERIMETER = 12" + 12" + 17" + 17"

PERIMETER = 58 inches

17"

12"

Notice that area is expressed in square units while the perimeter is expressed in linear units.

SEE ALSO: AREA OF A SQUARE, AREA OF A TRIANGLE

AREA OF A SQUARE = S^2

The area of a square is given by S^2 where S is the measure of one side. Since area is length x width, the area of a square is side x side, said as "side squared."

Example: What is the area of a square with a side of six feet?

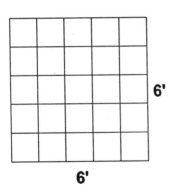

6'

6'

The area is 6' x 6' = 36 square ft. (sq ft). The units of area may also be expressed 36 ft^2.

It is important to carry along the units and not just the numbers when doing these problems.

Area is a two-dimensional measure.

CONTINUED NEXT PAGE

A square is always a rectangle. This is so because a square meets the definition of a rectangle. The opposite sides of a square are equal and the angles making up the figure are right angles (90 degrees). However, not all rectangles are squares. What makes a square a square is the fact that all four of its sides are equal and the angles that make up the square are right angles.

SEE ALSO: AREA OF A RECTANGLE

AREA OF A TRIANGLE

The area if a triangle is given by the formula: $A = 1/2\ B \times H$, read as: "one-half the base times the height."

This formula doesn't have to be memorized. If you realize that a triangle is half a rectangle, and the area of a rectangle is length times width, just take 1/2 of the area of a rectangle that has a length equal to the height and a width equal to the base of the triangle.

Example: Calculate the area of a triangle having a base of 4" and a height of 7".

Formula: $A = 1/2\ (B \times H)$

$A = 1/2\ (4" \times 7") = 1/2(28\text{ sq in}) = 14\text{ in}^2$

The answer is read: "fourteen square inches" but *NOT* "fourteen inches squared," (which might produce a wrong answer of 196 sq in).

CONTINUED NEXT PAGE

The rectangle and triangle would look like this:

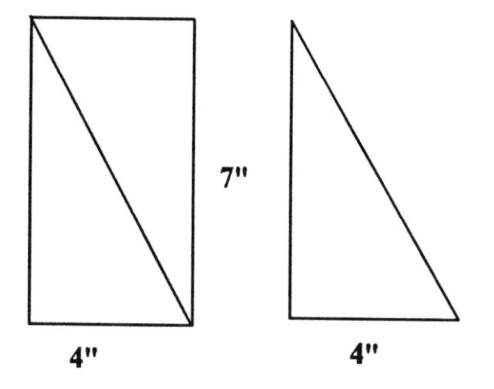

7"

4"   4"

The area of the triangle is one-half that of the rectangle.

SEE ALSO: AREA OF A RECTANGLE

ASSOCIATION

The association rule is one of the most important in mathematics. The word *associate* means to group. Sometimes numbers may be grouped differently and sometimes they may not.

Examples: $(3 + 4) + 5 = 3 + (4 + 5)$ These associations are
                $12 = 12$ OK. They produce true
                                              statements.

                 $(3 \times 4) \times 5 = 3 \times (4 \times 5)$
                     $60 = 60$

Examples: $(5 - 2) - 1 \neq 5 - (2 - 1)$ These associations are
                 $2 \neq 4$ not OK. They
                                              produce untrue
                 $(12 \div 3) \div 2 \neq 12 \div (3 \div 2)$ statements.
                     $2 \neq 8$

The rules of association are:

You may associate with respect to addition and multiplication.
You may NOT associate with respect to subtraction and division.

A similar companion rule is that of COMMUTATION.

You may commute with respect to addition and multiplication.
You may NOT commute with respect to subtraction and division.

SEE ALSO: COMMUTATION

AVERAGE (MEAN)

Symbol: $\overline{X}$

Definition: The average is the sum of the values of a given number of items divided by the number of items.

The *average* is a measure of central tendency. Differences from the average indicate how "far out" a value is. The greater the number of items, the more representative is the average.

Example: What is the average of the grades 64, 72, 84, 96.

Step 1. Add the values: $64 + 72 + 84 + 96 = 316$

Step 2. Divide the sum of the values by 4.

$$316 \div 4 = 79$$

The *average* is a statistic used to report rainfall, temperatures, prices, annual changes. Another measure similar to the average is the *median* which is the *middle* value. Put the values in order from smaller to largest and count to the middle value. This value is the *median.*

CONTINUED NEXT PAGE

Example: Find the *median* of the following scores:

36, 39, 54, 19, 28, 56, 72.

Step 1. Write the scores in order from lowest to highest.

19, 28, 36, 39, 54, 56, 72.

Step 2. Count in to the middle score which is 39.

When there is an *even* number of scores, the *median* is the average of the two middle scores.

Note that when calculating a mean, the division may not come out evenly. Simply perform decimal division to two places and round off to one place.

Example: The average of 36, 33, 93, 67, 57 = 57.2

SEE ALSO: DECIMALS - DIVISION

CIRCLES - EXAMPLES

Example: A circle has a radius of 6 inches. Find its diameter, circumference and area.

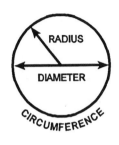

D = 2R $D = 2 \times 6" = 12"$

C = D x pi $C = 12" \times 3.14 = 37.68"$

$C = 12" \times 22/7 = 37.71"$

A = pi x R² $A = 3.14 \times (6")^2 = 113.04 \text{ sq in}$

$A = 22/7 \times (6")^2 = 113.14 \text{ sq in}$

The differences in circumference and area calculations is the result of choosing the fractional or decimal approximations of pi.

When 22/7 is used as an approximation of pi, its decimal value comes out 3.14286. The decimal value of pi to five places is 3.14159.

CONTINUED NEXT PAGE   **19**

This difference will produce slightly different final answers when pi is a factor in the calculations.

Another difference occurs when the value of pi is taken from a calculator. Most calculators provide a value of pi to 9 places. For example: 3.141592654. All forms are correct.

So, what do you do? The decimal value for pi is more precise than the fractional approximation. However, *all three* produce 3.14 when rounded to two places, so there really is not much difference unless you are doing very precise calculations. To be safe, specify whether you are using 22/7 or 3.14 or the value of pi from the calculator.

SEE ALSO: CIRCLES - TERMS - AREA

20

CIRCLES - TERMS - AREA

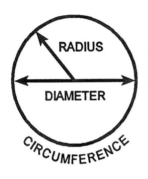

R = RADIUS - the distance from the center of the circle to any point on its circumference.

D = DIAMETER - the distance across the circle through the center.

C = CIRCUMFERENCE - the distance around the rim of the circle.

pi $\cong$ 3.14 . . . $\cong$ 22/7
(these are approximations)

pi = the number of diameters that will wrap around one circumference

1 diameter	1 diameter	1 diameter	0.14D

C = D x pi   D = 2R   A = pi x R^2

CONTINUED NEXT PAGE   21

THE VALUE OF pi

Pi is the number of times that the diameter of a given circle will fit into the circumference. The diameter wraps around three full times and a "little bit." That little bit is 0.14. When added to the three times that the diameter *does fit* around the circumference the number comes out 3.14.

The value of pi gave the ancient mathematicians a great deal of difficulty. To the Greeks, pi was an insult to mathematics because it made the circle "imperfect" in their eyes. If the diameter was twice the radius (a simple ratio) then why didn't the diameter fit exactly three times around the circumference?

Pi is a particularly mysterious number because it has no ending. It does not repeat and it does not terminate like other decimals. For this reason, pi cannot be expressed as a fraction. It is not a *rational* number. 22/7 is an *approximation* of the value of pi and may be used without fear of error. With modern computers pi has been calculated out to a very large number of decimal places.

SEE ALSO: CIRCLES - EXAMPLES

COMMON DENOMINATOR

A common denominator is needed to add and subtract fractions with different denominators such as 1/2 + 1/4.

Definition: A common denominator is a whole number that is evenly divisible by all the denominators in the problem.

How to get a CD: Multiply all the different denominators together. *The product of all the denominators is a common denominator because all the factors HAVE TO divide evenly into the product.*

Example: Find a CD for 1/2, 1/3, 1/4, 1/5.

Answer: 2 x 3 x 4 x 5 = 120 which is a common denominator.

Note: 120 is NOT the LCD (Least Common Denominator) but it is a common denominator and will work in all cases.

CONTINUED NEXT PAGE   23

There is nothing magical about the least common denominator. Any common denominator will work when adding or subtracting fractions. However, it often is less cumbersome to find a least common denominator.

A method to find the LCD is found under LEAST COMMON DENOMINATOR.

SEE ALSO: LEAST COMMON DENOMINATOR

COMMUTATION

The commutation rule is one of the most important in mathematics. The word *commute* means to *move*. Sometimes numbers may be moved around and sometimes they may not.

Examples: $3 + 5 = 5 + 3$ These commutations are OK.
              $8 = 8$ They produce true statements.

             $8 \times 7 = 7 \times 8$
              $56 = 56$

Examples: $8 - 3 \neq 3 - 8$ These commutations are not OK.
              $5 \neq -5$ They produce untrue statements.

             $12 \div 4 \neq 4 \div 12$
              $3 \neq 1/3$

The rules of commutation are:

You may commute with respect to addition and multiplication
You may NOT commute with respect to subtraction and division.

A similar companion rule is that of ASSOCIATION.

You may associate with respect to addition and multiplication.
You may NOT associate with respect to subtraction and division.

SEE ALSO: ASSOCIATION

COMPARING FRACTIONS

Comparing fractions means to determine which fraction is larger than another. Often the problem requires you to list the fractions *in order* from smallest to largest, or largest to smallest.

Example: Write in order from smallest to largest:

$$\frac{3}{4} \quad \frac{2}{3} \quad \frac{5}{6} \quad \frac{1}{2}$$

Step 1.   Find a common denominator (12 is a good CD).

Step 2.   Write equivalent fractions

$$\frac{9}{12} \quad \frac{8}{12} \quad \frac{10}{12} \quad \frac{6}{12}$$

Step 3.   Write fractions in order by numerator-- smallest to largest.

$$\frac{6}{12} \quad \frac{8}{12} \quad \frac{9}{12} \quad \frac{10}{12}$$

CONTINUED NEXT PAGE

ALTERNATE METHOD

Step 1.   Convert all the fractions into their equivalent decimals by
          dividing the numerator by the denominator.

3/4	=	0.75
2/3	=	0.666 . . .
5/6	=	0.833 . . .
1/2	=	0.5

Step 2.   Place decimals in desired order: 0.5, 0.666, 0.75, 0.833

Step 3.   Write equivalent fractions in same order:

1/2, 2/3, 3/4, 5/6

SEE ALSO: EQUIVALENT FRACTIONS,
           FRACTIONS TO DECIMALS   28

COMPLEX FRACTIONS

A *complex fraction* is one where the numerator or the denominator of a fraction is a fraction itself.

Examples:
$$\dfrac{\dfrac{1}{2}}{\dfrac{3}{4}} \qquad \dfrac{1}{\dfrac{1}{2}} \qquad \dfrac{1\,\dfrac{1}{2}}{\dfrac{2}{5}}$$

The way to treat complex fractions is to look at them as divisions. Divide the fraction in the numerator by the fraction in the denominator.

For example, the first complex fraction can be re-written as:

$$\frac{1}{2} + \frac{3}{4} - \frac{4}{6} - \frac{2}{3}$$

which is nothing but a fraction divided by a fraction.

CONTINUED NEXT PAGE

The numerator or the denominator might also be a mixed number. The third problem is such an example.

Step 1.   Change the mixed number into the improper fraction 3/2

Step 2.   Set up as a division of fractions:

$$\frac{3}{2} \div \frac{2}{5} \cdot \frac{15}{4}$$

SEE ALSO: EQUIVALENT FRACTIONS, FRACTIONS - DIVISION, RECIPROCALS

COMPOSITE NUMBERS

Whole numbers are either *prime* or *composite.* (Zero or 1 is neither.)
Composite numbers are products of two or more whole numbers other
than 1 or zero.

Composite numbers are resolved into their prime factors by using the
FACTOR TREE.

Example: 24 is "composed" of 2 x 2 x 2 x 3, all of which are prime
factors. (A factor is simply a multiplier.)

A *prime number* is one which is divisible only by one or itself.

SEE ALSO: FACTOR TREE, PRIME NUMBERS   **31**

CONVERSION RATIOS

16 ounces	one pound	60 seconds	one minute
2.2 pounds	one kilogram	60 minutes	one hour
1000 grams	one kilogram	24 hours	one day
2000 pounds	one ton	7 days	one week
32 ounces (liquid)	one quart	28 days	one lunar month
2 pints	one quart	365 days	one standard year
4 quarts	one gallon	366 days	one leap year
55 gallons	one barrel	10 years	one decade
1.06 quarts	one liter	100 years	one century
12 inches	one foot	12 items	one dozen
36 inches	one yard	12 dozen	one gross
3 feet	one yard	1000 millirems	one REM
5,280 feet	one mile		
2.54 centimeters	one inch		
100 centimeters	one meter		**METRIC VOLUMES**
1000 millimeters	one meter		
144 sq inches	one square foot	1 cubic cm	one milliliter
1,728 cu inches	one cubic foot	1000 ml	one liter
27 cubic feet	one cubic yard	100 cubic ml	one decaliter
		1 cubic meter	1,000,000 cc

CONTINUED NEXT PAGE

Note that these conversion ratios are all equal to ONE. 16 ounces IS one pound, therefore you can multiply any quantity by any of these ratios and not change the value of the quantity.

Note that the ratios can be read *forwards* and *backwards,* i.e., 16 ounces = one pound and one pound = 16 ounces.

Units should always be carried along with the numerical parts. The units can be treated mathematically. If the same unit appears in a denominator and a numerator, the units divide out.

Example: $36 \text{ INCHES} \times \dfrac{1 \text{ FOOT}}{12 \text{ INCHES}} = 3 \text{ FEET}$

This method of *unit analysis* is valuable in medical, technical, and scientific work. It is a skill worth developing.

SEE ALSO: METRIC SYSTEM

CROSS MULTIPLICATION

It is easy to confuse "cross multiplication" and "multiplying across."

CROSS MULTIPLICATION refers to a method of dealing with proportions and equivalent fractions when one of the numbers is missing. The *cross* refers to the X pattern of multiplication. The numerator of one fraction is multiplied by the denominator of the other.

In order to *cross multiply* it is necessary to have an equal sign between two fractions. Cross multiplication is NOT used when multiplying fractions.

Example: Find the missing value: $\dfrac{2}{3} = \dfrac{?}{6}$

Step 1.   Multiply 2 x 6 = 12 (cross product).

Step 2.   Divide 12 by 3.

             Missing value is 4.

CONTINUED NEXT PAGE

Rule: *If two fractions are equivalent, their cross products are equal.*

Example: Given the equivalent fractions:

$$\frac{2}{3} = \frac{6}{9}$$

note that: $\quad 2 \times 9 = 3 \times 6$

2×9 is one cross product; 3×6 is the other. Both products are equal to 18.

MULTIPLYING ACROSS refers to the operation of multiplying fractions.

Rule: *To multiply two fractions, multiply their numerators; multiply their denominators.*

Example: $\quad \dfrac{4}{5} \times \dfrac{5}{7} = \dfrac{20}{35}$

SEE ALSO: EQUIVALENT FRACTIONS,
FRACTIONS - MULTIPLICATION

CUBES AND CUBE ROOTS

The *cube* of a number is the number that results when a number is used as a factor three times.

Examples: The cube of 2 is: $2 \times 2 \times 2 = 8$

The cube of 3 is: $3 \times 3 \times 3 = 27$

The cube of 4.1 is: $4.1 \times 4.1 \times 4.1 = 68.921$

Another way to express a cube is to raise the number to the third power.

Examples: $2^3 = 8$; $3^3 = 27$; $(4.1)^3 = 68.921$

Cubic measure is three dimensional, such as VOLUME. Some cubic units are: *cubic* feet, *cubic* centimeters (cc), *cubic* miles. These units may also be written ft^3, cm^3, mi^3. The length, width and height of a cube have the same numerical measure, so the volume of a cube is S^3 read as "side cubed."

CONTINUED NEXT PAGE

The *cube root* of a number is just the opposite of cubing. It is the value of one *factor* (multiplier) that produces the number,

Examples: If 2 cubed is 8, the cube root of 8 is 2.
              If 3 cubed is 27, the cube root of 27 is 3.

There are several ways to find the cube root of any number.

1.   Use a table of cube roots (SEE CUBE ROOTS - TABLE)

2.   Use a calculator. The cube root key looks like: $\sqrt[3]{x}$

3.   Memorize the first five perfect cubes:

$$1^3 = 1$$
$$2^3 = 8$$
$$3^3 = 27$$
$$4^3 = 64$$
$$5^3 = 125$$

The cubes and cube roots of negative numbers are negative.

SEE ALSO: POWERS OF NUMBERS, VOLUME

38

CUBE ROOTS - Table of Cubes and Cube Roots

No	Cube	Cube Root		No	Cube	Cube Root
1	1	1		10	1000	2.15
2	8	1.26		11	1331	2.22
3	27	1.44		12	1728	2.29
4	64	1.59		13	2197	2.35
5	125	1.71		14	2744	2.41
6	216	1.82		15	3375	2.47
7	343	1.91		16	4096	2.52
8	512	2		17	4913	2.57
9	729	2.08		18	5832	2.62

CONTINUED NEXT PAGE

No	Cube	Cube Root		No	Cube	Cube Root
19	6859	2.67		28	21952	3.04
20	8000	2.71		29	24389	3.07
21	9261	2.76		30	27000	3.11
22	10648	2.80		31	29791	3.14
23	12167	2.84		32	32768	3.17
24	13824	2.88		33	35937	3.21
25	15625	2.92		34	39304	3.24
26	17576	2.96		35	42875	3.27
27	19683	3		36	46656	3.30

You may use a calculator to find the cube or the cube root of a number.

DECIMALS - ADDITION AND SUBTRACTION

Rules: LINE UP THE DECIMAL POINTS. If a whole number is being added, its decimal point is to the right of the last digit including any zeroes.

Hint: Use graph paper or ruled paper turned sideways as an aid to keeping the decimal places in line.

Example: Add 5.67 + 3 + 0.007

Proper set-up:

$$
\begin{array}{r}
5.67 \\
3 \\
\underline{0.007} \\
8.677
\end{array}
$$

If a decimal is subtracted from a whole number, LINE UP THE DECIMAL POINTS. Add as many zeroes to the right of the decimal point as needed, then perform the subtraction.

CONTINUED NEXT PAGE

Example: Perform the indicated subtraction: 5 - 0.067

Line up   5.000   Zeroes added to 3 places
decimal   - 0.067
points   4.933

When using a calculator to do subtractions, use the subtraction key
and not the +/- key. The +/- key is used to change the SIGN of the
number.

Put 5 into the calculator first, then depress the subtraction button, then
put in 0.067 and then press the = button. It is not necessary to input the
5 as 5.000

DECIMALS - DIVISION

Rule: The divisor must be a whole number. To change a decimal divisor to a whole number, the decimal point must be "moved." See next page for an explanation of why decimal points are "moved."

Example: 10.1 $\overline{\smash{\big)}\ 40.56}$

Step 1. Move decimal of divisor one place to the right and move the decimal of the dividend one place to the right.

101 $\overline{\smash{\big)}\ 405.6}$

Now the divisor is a whole number.

Step 2. Place a decimal directly above the line.

101 $\overline{\smash{\big)}\ 405.6}$

CONTINUED NEXT PAGE   43

Step 3. Do the division as you would with whole numbers. Carry out the division as many places as requested. If the number of places is not requested, carry out to three decimal places and round off to two.

$$4.02$$

$$101 \overline{)\ 405.60}$$ A zero has been added to the dividend.

The decimal point is not magically "moved" right or left at whim. In division the reason is quite simple. A division may be expressed as a fraction. Expressing the long division as a fraction yields:

$$\frac{40.56}{10.1}$$

If the numerator and the denominator are multiplied by 10, the result is:

$$\frac{405.6}{101}$$

producing a whole number in the denominator (which is the divisor).

SEE ALSO: POWERS OF 10, ROUNDING DECIMALS

DECIMALS TO FRACTIONS

To change a decimal to a fraction, READ the decimal and WRITE it as you hear it. YOU NEED TO KNOW THE *PLACE NAMES* OF DECIMALS.

Example: Write 0.24 as a fraction.

0.24 is read as "twenty-four hundredths." Note the word *hundredths.* *Hundredths* is a fraction with a denominator equal to 100. Therefore, $0.24 = 24/100$

Example: Write 0.548 as a fraction.

0.548 is read as "five hundred forty-eight thousandths.   The denominator is 1000. So, $0.548 = 548/1000$

What about simple fractions like 1/4?   How are they related to decimals?

If you divide the numerator (1) by the denominator (4) the result is 0.25 And 0.25 is read as twenty-five hundredths. $25/100 = 1/4$

CONTINUED NEXT PAGE   45

What about mixed numbers? How can mixed numbers be converted to decimals?

Use the same system. *READ* the mixed number and *WRITE* it, remembering that a mixed number is the addition of a whole number and a fraction.

Example: Write 1.28 as a mixed number.

Read 1.28 as "one *AND* twenty-eight hundredths. Now write the mixed number as:

$$1 \frac{28}{100}$$

SEE ALSO: FRACTIONS TO DECIMALS
            PLACE NAMES OF NUMBERS

DECIMALS - MULTIPLICATION

Rules: DO NOT try to line up the decimal points. IGNORE the decimal points. (It would help to re-write the problem *without* the decimal points.)

Step 1.   Multiply the whole numbers together.

Step 2.   Count the *total* number of decimal places in both of the factors. (Factors are the multipliers.)

Step 3.   Move the decimal point from right to the left, the number of places you counted in step 2.

Example: Multiply 3.47 x 2.14 (there are four decimal places here).

Step 1.   347 x 214 = 74258

Step 2.   There are four decimal places.

CONTINUED NEXT PAGE   **47**

Step 3. Starting to the right of the 8, move the decimal four places to the left.

7.4258

Note: It is customary to round off to the smallest number of decimal places in any one of the factors. This is so because it is not possible to have a product *more precise* than the least precise factor. The least precise factor has a precision of hundredths. The final answer should be precise to two decimal places.

The final answer would be 7.43

SEE ALSO: PLACE NAMES OF NUMBERS

DECIMALS TO PERCENTS

Rule: To change a decimal to a percent, multiply the decimal by 100%.

Example: Change 0.16 into a percent

0.16 x 100% = 16%

Note: 100% = 1 because the % sign means to divide by 100, so

100% = 100/100 = 1

There is no difference in value between 16%, 0.16 or 16/100

Example: Change 14 into a percent

14 x 100% = 1400%

Although 1400% may look strange at first, it means a 14 fold change.

CONTINUED NEXT PAGE

49

Example: Change 0.005 into a percent.

$$0.005 \times 100\% = 0.5\%$$

This too may look strange, but it means one-half of a percent.

Note: To say that you multiply by 100 is not correct. You cannot multiply a number by 100 without changing its value. Since 100% = 1, you CAN multiply by 100% without changing the value of a number. That is why:

$$0.16 = 16/100 = 16\%$$

SEE ALSO: DECIMALS TO FRACTIONS,
FRACTIONS TO DECIMALS,
PERCENTS TO DECIMALS

DECIMALS vs FRACTIONS

There may be a difference in final answers depending whether decimals or fractions are used in calculating. Some fractions convert evenly into equivalent decimals such as 1/4 = 0.25. However, many fractions *do not* convert evenly, such as 1/7 = 0.1428571 ... or 1/3 = 0.333 ....

A particular difficulty arises in the use of pi in problems. Pi = 22/7 or 3.14159 .... Both of these values are *approximations.*

To handle repeating and non-repeating decimals, round off the answer to the second place to the right of the decimal unless instructed otherwise. DO NOT round off any decimals before the final answer is calculated. Errors will creep into the calculations if this is done. *When using a calculator, keep all the places displayed till the very last calculation, then round off.*

Should you use fractions or decimals? Use whatever form feels comfortable and don't be afraid of converting one into the other.

CONTINUED NEXT PAGE

In a whole number the decimal point is not normally shown, but it really exists. It is found to the right of the last digit (including any zeroes).

Example: In the number 586, the decimal is at the right end of 586.

             In the number 4600, the decimal is to the right of the last zero, 4600.

             In the number 100, the decimal is to the right of the last zero, 100.

Note: It is not wrong to place a decimal at the end of a whole number. If it helps you in calculating, then just do it.

Note: The decimal system is not symmetrical. There are ONES to the left of the decimal point, but there are no ONES to the right of the decimal point. There are TENTHS to the right of the decimal, but no TENTHS to the left.

SEE ALSO: FRACTIONS TO DECIMALS,
          PLACE NAMES OF NUMBERS   52

DIVISION FACTS

A division may always be expressed as a fraction.
In a fraction, the denominator is the *divisor.*

ALWAYS divide by the denominator.

ALWAYS divide by the number that comes *after* the ÷ sign.

$$\overset{\text{QUOTIENT}}{\text{DIVISOR} \,\big/\, \text{DIVIDEND}}$$

The *divisor* is the number you are dividing by. When using a calculator, the divisor goes in *after* the ÷ sign.

The *dividend* is the number you are dividing into.

The *quotient* is the answer.

The divisor does not have to be smaller than the dividend. It is a common error to think that the smaller number always divides into the

CONTINUED NEXT PAGE

larger number. *Any number may be divided by any other number except division by zero is not allowed. NEVER divide by zero.*

Example: $12 \div 14$ means 12 / 14 or 12 divided by 14 or

$$14 \overline{\smash{\big)}\ 12}$$

When doing this problem on a calculator, first input the 12, then the division sign, then the 14 and then the = sign button.

Answer: 0.8571428

Which may be rounded off to 0.86

EQUIVALENT FRACTIONS

Equivalent fractions are fractions that are equal in value.

Example: $\dfrac{4}{5} = \dfrac{8}{10}$ (numerator and denominator
were multiplied by 2)

Any fraction may be "changed" into *many* equivalent fractions by multiplying the numerator and denominator by the same quantity.

Examples: $\dfrac{1}{2} = \dfrac{2}{4} = \dfrac{3}{6} = \dfrac{4}{8} = \dfrac{5}{10}$ . . .

Another way to think about equivalency is to multiply the fraction by 1 expressed as 2/2 or 3/3 or 4/4 or any fraction equivalent to 1. Any number divided by itself is 1. (n/n)

Uses: Equivalent fractions are useful in finding the "missing" value in proportions, simplifying fractions, adding or subtracting fractions.

CONTINUED NEXT PAGE   55

Example: Use the concept of an equivalent fraction to find the missing value in the following proportion.

$$\frac{5}{8} = \frac{?}{32}$$

Step 1. Ask yourself: "What number times 8 produces 32?" Answer: 4.

Step 2. Multiply 5/8 by 4/4 to get the "missing" number 20.

$$\frac{4}{4} \times \frac{5}{8} = \frac{20}{32}$$

SEE ALSO: PROPORTIONS

FACTORS - Prime and Otherwise

When two numbers are multiplied together they form a *PRODUCT*. The numbers multiplied together are called *FACTORS*. A given product may have many sets of factors. In the equation:

$3 \times 4 = 12,$ 3 and 4 are *factors* of 12.

If a factor cannot be divided by any number other than 1 or itself, the factor is said to be a *PRIME FACTOR*. Prime numbers may be found in a PRIME NUMBERS TABLE.

The first 10 prime numbers are 2, 3, 5, 7, 11, 13, 17, 19, 23, 29.

Numbers which can be divided evenly by numbers other than 1 or the number itself are called *COMPOSITE*. 12 is composite because it is "composed" of other factors and can be divided by 2, 3, 4 and 6.

Notice that some numbers may be composed of more than one set of factors, but that there is only ONE set of PRIME FACTORS for a given composite number.

CONTINUED NEXT PAGE   57

The easiest way to determine if a number is a prime number is to look it up on a PRIME NUMBERS TABLE. Also, if a number appears in the product section of a MULTIPLICATION TABLE, that number *cannot* be prime because it is a product of two other numbers.

The PRIME FACTORS of a composite number can be found by using a technique called the FACTOR TREE.

SEE ALSO: FACTOR TREE, PRIME NUMBERS,
                PRIME NUMBER TABLE

FACTOR TREE

A simple and effective way to find all the prime factors of a composite number is to use a FACTOR TREE.

Step 1. Find ANY factor that divides evenly into the composite number.

Step 2. Write down the divisor and the quotient. Use the prime number table to see if either is PRIME. Circle the primes.

Step 3. Continue dividing all composite factors till all quotients are prime.

Step 4. Arrange the circled prime factors in order. The prime factors of 120 are:

2 x 2 x 2 x 3 x 5 OR 2^3 x 3 x 5

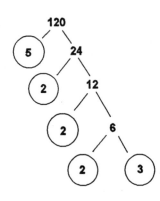

CONTINUED NEXT PAGE

The following example demonstrates that it is not necessary to use prime divisors for every division -- the tree will simply have more branches.

Notice that the result is always the same.

The prime factors of 120 are:

2 x 2 x 2 x 3 x 5

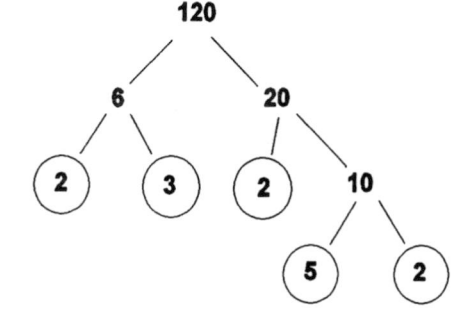

SEE ALSO: FACTORS, PRIME NUMBER TABLE

FORWARDS AND BACKWARDS

Going "backwards" is a math skill worth developing. For example, there are 12 inches in one foot, but it is also true that one foot = 12 inches. But how many feet are there in one inch? Ans: One-twelfth of a foot in one inch.

Subtraction and addition are operations that are the reverse of each other. Multiplication is the reverse of division and division is the reverse operation of multiplication.

The concept of a reciprocal is also a "backwards" kind of thinking.

Multiplying by 1/2 is the same as dividing by 2.

When you multiply by a fraction you are also dividing by the reciprocal of the fraction.

Questions like "What number multiplied by 6 is equal to 30?" or, "18 is what % of 48?" use "backwards" reasoning skills.

CONTINUED NEXT PAGE   61

Actually, there is nothing "backwards" about such notions. The reason why they may look strange is that math skills are usually taught only in one direction. Math facts such as 4 x 6 = 24 should also be thought of in the form: 24 is the product of 6 and some number.

Backwards thinking is also useful when dealing with CONVERSION FACTORS.

Example: 1 liter = 1.06 quarts   AND   1.06 quarts = 1 liter

If there are 1.06 quarts per liter, then there must be one liter per 1.06 quarts and the conversion ratios may be written:

$$\frac{1 \text{ liter}}{1.06 \text{ quarts}} = \frac{1.06 \text{ quarts}}{1 \text{ liter}} = 1$$

The two different forms make it easy to convert one unit into another by multiplying by the appropriate fraction. And since both fractions are equal to 1, there is no change in the value of the expression when you multiply, just different equivalent units.

SEE ALSO: CONVERSION FACTORS   62

FRACTIONS - ADDITION

Rule: If the denominators are *alike*, just add the numerators and keep the same denominator.

Example: $3/5 + 1/5 = 4/5$

Example: $5/7 + 6/7 = 11/7$ (it is OK to leave in this form)

Rule: If the denominators are *unlike*:

Step 1. Find any CD (Common Denominator).

Step 2. Write equivalent fractions for each fraction.

Step 3. Add the numerators.

Step 4. Keep the common denominator.

CONTINUED NEXT PAGE

Example: Add the fractions 1/2, 1/3, 1/4

Step 1. One common denominator for 2, 3 and 4 is 12

Step 2. Equivalent fractions are:

$$1/2 = 6/12$$

$$1/3 = 4/12$$

$$1/4 = 3/12$$

Step 3.
Step 4. Ans: 13/12

If you wish to change the improper fraction into a mixed number, divide the numerator by the denominator and write the remainder over the denominator.

$$\frac{13}{12} = 1\frac{1}{12}$$

SEE ALSO: EQUIVALENT FRACTIONS

FRACTIONS - DEFINITION

A FRACTION IS A DIVISION

The NUMERATOR is always divided by the DENOMINATOR.

TERMS: nUmerator (U for Up)   Denominator (D for Down)

ALWAYS DIVIDE BY THE DENOMINATOR.

The line between the N and D is a *division* line.

Whole numbers may be expressed as fractions by dividing by 1

Example: $14 = \dfrac{14}{1}$

CONTINUED NEXT PAGE

The definition of a fraction is perhaps the most neglected concept in mathematics education. From the definition that a fraction is a division flows a list of important skills:

- Change a fraction to a decimal by dividing the numerator by the denominator.

- Ratios are fractions.

- A proportion is a pair of two equivalent fractions.

- Reciprocal pairs are written as fractions.

SEE ALSO: FRACTIONS TO DECIMALS,
PROPORTIONS,
RATIO,
RECIPROCALS

FRACTIONS - DIVISION

YOU DO NOT NEED A CD TO DIVIDE FRACTIONS

Method: Multiply the first fraction by the reciprocal of the second.

Example: Divide $\dfrac{3}{4} \div \dfrac{1}{2}$

Write as: $\dfrac{3}{4} \times \dfrac{2}{1} = \dfrac{6}{4}$

Simplify as desired to:

$$\dfrac{3}{2} \quad \text{OR} \quad 1\dfrac{1}{2}$$

Unless required to do so, 6/4 is a perfectly good answer. However, many math tests require simplest terms for correct answers.

CONTINUED NEXT PAGE

The method shown on the last page is the rule of thumb called "invert and multiply" and may be used if understood. The word "invert" does not accurately describe what is happening to the fraction. The correct words used to describe division of fractions uses the reciprocal notion.

But why does the "invert and multiply" rule work? Let's do the problem again.

Step 1: Express the division as a complex fraction.

Step 2. Multiply the numerator and denominator by the reciprocal of the denominator.

$$\frac{\dfrac{3}{4}}{\dfrac{1}{2}} \times \frac{\dfrac{2}{1}}{\dfrac{2}{1}} = \frac{\dfrac{6}{4}}{\dfrac{1}{1}} = \frac{3}{2}$$

SEE ALSO: RECIPROCAL

FRACTIONS - MULTIPLICATION

YOU DO NOT NEED A CD TO MULTIPLY FRACTIONS

Rule: Multiply numerators, multiply denominators. Simplify to lowest terms.

Example: $\dfrac{3}{5} \times \dfrac{4}{9} = \dfrac{12}{45} = \dfrac{4}{15}$

It does not matter how many fractions there are in the problem. Just multiply all the numerators together and all the denominators together. As a last step, simplify to lowest terms.

Multiplication of fractions can be better understood if the x of multiplication is changed to the word "of." For example, 1/2 x 1/2 means more when expressed: "What is 1/2 of 1/2?" The answer, 1/4 is easier to to see than "timesing" 1/2 by 1/2.

At times it is possible to simplify a multiplication.

CONTINUED NEXT PAGE

Use this important rule of fractions:

You may multiply or divide the numerator and the denominator of any fraction by the same quantity without changing the value of the fraction.

Example: Multiply

$$\frac{5}{8} \times \frac{8}{9} \times \frac{9}{10} = \frac{5}{10} = \frac{1}{2}$$

Notice that the denominator of 5/8 will divide once into the numerator of 8/9 and the denominator of 8/9 will divide once into the numerator of 9/10. This process is erroneously referred to as *cancelling*. Nothing actually gets *cancelled*. There is always a quotient, even if it is 1 which is not often written down.

The numbers that divide out do not have to be next to one another. Any numerator or denominator will do as long as they are divisible by the same quantity.   This procedure is possible because of the COMMUTATION rule with respect to multiplication.

SEE ALSO: COMMUTATION   70

FRACTIONS SUBTRACTION

Rule: If the fractions have the same denominator, subtract the numerators. Write the difference over the denominator.

Example: $\dfrac{5}{6} - \dfrac{2}{6} = \dfrac{3}{6}$ OR $\dfrac{1}{2}$

Rules: If the denominators are different

Step 1.   Find a common denominator (CD).

Example: $\dfrac{2}{3} - \dfrac{4}{9}$ Common denominator is 9

Step 2.   Write equivalent fractions using the CD.

$$\frac{2}{3} = \frac{6}{9} \quad \text{and} \quad \frac{4}{9} = \frac{4}{9}$$

CONTINUED NEXT PAGE

Step 3. Subtract the numerators

$$6 - 4 = 2$$

Step 4. Write the difference over the common denominator

$$\frac{2}{9}$$

If the second numerator is larger than the first numerator, the answer will come out negative.

Example: $\dfrac{3}{7} - \dfrac{5}{7} = -\dfrac{2}{7}$

Sometimes the resulting fraction may be simplified to lowest terms.

SEE ALSO: EQUIVALENT FRACTIONS

72

FRACTIONS TO DECIMALS

To change a fraction into a decimal, divide the numerator by the denominator.

There are two ways in which the result may appear:

-- the decimal will come out even and stop
-- the decimal will repeat

Examples: 1/8 = 0.125 (terminal)
             1/3 = 0.3333333 (repeating)
             1/7 = 0.142857142857 (repeating)

With the repeating decimals, round off to the second place to the right of the decimal point.

Example: 1/7 = 0.14

Some fractions have rather long repetition groups and the repetition may not be noticed on the calculator.

CONTINUED NEXT PAGE

The common fractions have common decimals that should become familiar with use such as 1/4 = 0.25, 1/2 = 0.5

An alternative method of expressing repeating decimals is to put a bar over the part that repeats.

Examples: 1/3 = $0.\overline{33}$

1/7 = $0.\overline{142857}$

SEE ALSO: DECIMALS TO FRACTIONS,
                      ROUNDING DECIMALS

FRACTIONS TO PERCENTS

Rule: To change a fraction to a percent, first change the fraction to a decimal, then change the decimal to a percent.

Example: Change 3/5 into a percent.

3/5 = 0.6

0.6 x 100% = 60%

Notice that the decimal is multiplied by 100%, NOT just by 100.
100% = 1

Example: Change 1 into a percent.

1 x 100% = 100%

Example: Change 0.5 into a percent.

0.5 x 100% = 50%

CONTINUED NEXT PAGE

When you have the WHOLE THING you have 100% of it. When you have ONE of something you have 100% of it.

$$1 = 1/1 = 100\% = 100/100$$

Some numbers look rather strange, but the rules are the same.

Example: Change 14 into a percent.

$$14 \times 100\% = 1400\%$$

Fourteen hundred percent is perfectly OK. It means that there is a fourteen-fold change in some number.

Example: Change 0.002 into a percent.

$$0.002 \times 100\% = 0.2\%$$

This answer is read as "two-tenths of a percent."

SEE ALSO: FRACTIONS TO DECIMALS, PERCENT

GEOMETRIC SHAPES

TRIANGLES are three-sided enclosed figures. The sides may or may not be equal. The angles may or may not be equal. Each different triangle has its own name depending on its properties. SEE TRIANGLES for more information.

SQUARES are four-sided closed figures whose sides are of equal size and all the angles are right angles (90°). Squares belong to a family of figures called parallelograms because their opposite sides are parallel and equal. Some other members of this family are rectangles and rhombuses. SEE SQUARES.

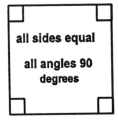

all sides equal

all angles 90 degrees

CONTINUED NEXT PAGE

PARALLELOGRAMS are four-sided figures whose opposite sides are equal and parallel. All squares, rectangles and rhombi are parallelograms.

RECTANGLES are parallelograms whose opposite sides are equal and whose angles are all right angles (90°). A square is a rectangle but not all rectangles are squares.

RHOMBI are diamond-shaped parallelograms. Their opposite sides are parallel and equal.

TRAPEZOIDS are four-sided figures that have only one set of opposite sides parallel. The opposite sides that are parallel cannot be equal but the non-parallel sides may be.

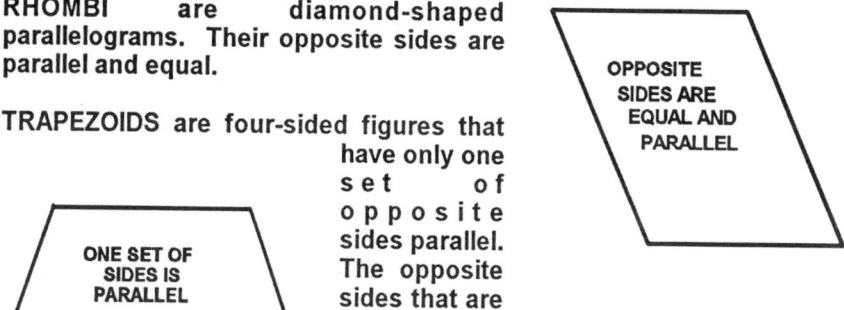

OPPOSITE
SIDES ARE
EQUAL AND
PARALLEL

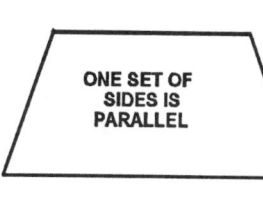

ONE SET OF
SIDES IS
PARALLEL

78

GEOMETRIC SHAPES II

PENTAGONS are five-sided figures. The sides and angles do not have to be equal, but if the sides AND angles are equal, the figure is called a *regular* pentagon. This term *regular* refers to all geometric forms.

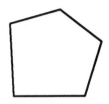

HEXAGONS are six-sided figures. As in pentagons, the sides do not have to be equal or parallel. The angles do not have to be equal either. If all the sides and angles of a hexagon are equal in length AND the angles are all equal, it is called a *regular* hexagon which has some special properties. The hexagon can be cut up into six equilateral triangles. These triangles have equal sides and equal angles and each of the angles is 60°. This symmetry is very helpful when doing geometrical calculations. A regular hexagon can be pictured as two identical trapezoids or three identical rhombi.

CONTINUED NEXT PAGE   79

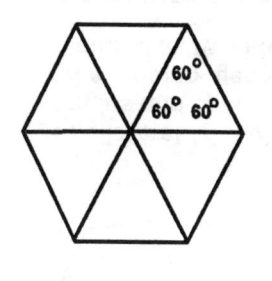

Each of the equilateral (equal-sided) triangles has three equal angles of 60° each. Since there are 120° in the two base angles of each triangle, and there are 6 such triangles, then the sum of the angles in a regular hexagon is 6 x 120° or 720°.

Hexagons are often found in nature such as in honeycombs and crystals. The hexagon is one of nature's strongest architectural forms because it will not collapse when squeezed like other forms.

Other geometric forms include OCTAGONS which are eight-sided figures. The *Stop Sign* is an octagon. The DECAGON is a ten-sided figure.

SEE ALSO: AREAS, SQUARES, TRIANGLES   80

GREATEST COMMON FACTOR

The *Greatest Common Factor* (GCF) is the largest whole number that will divide evenly into each of the numbers of the set. There are two conditions here. First, the GCF is *common* to all the numbers in question, and secondly, the number is the *largest* divisor possible.

Example: Find the GCF of 15, 30 and 60.

Look at the numbers. The GCF *cannot* be any larger than 15. It is also easy to see that 15 divides evenly into 15, 30 and 60. It is also the *largest* divisor.

In many problems the GCF is *not* evident by inspection, but the following method of PRIME FACTORS may be used.

Example: Find the GCF of 40, 50, 60, 120.

Step 1.   Find the *prime factors* of each number.

$$40 = 2 \times 2 \times 2 \times 5 = 2^3 \times 5^1$$

CONTINUED NEXT PAGE   81

$$50 = 2 \times 5 \times 5 = 2^1 \times 5^2$$

$$60 = 2 \times 2 \times 3 \times 5 = 2^2 \times 3^1 \times 5^1$$

$$120 = 2 \times 2 \times 2 \times 3 \times 5 = 2^3 \times 3^1 \times 5^1$$

Step 2.   Looking at the *exponential* forms, choose the *lowest* exponential of each *common* base.

Step 3.   Multiply these exponentials together.

$$2^1 \times 5^1 = 10 \quad \text{which is the GCF.}$$

Notice that the base 3 did not appear in the multiplication because 3 is NOT common to all four sets of factors, but the bases 2 and 5 are.

SEE ALSO: FACTOR TREE, POWERS OF NUMBERS

LEAST COMMON DENOMINATOR

The *Least Common Denominator* (LCD) is useful but not essential for adding or subtracting fractions. A COMMON DENOMINATOR works splendidly. The LCD is just a sophistication of arithmetic.

Definition: The LCD is the *smallest whole number* that is evenly divisible by all the denominators in the problem.

Method 1. INSPECTION. Look at the numbers and pick out the LCD. In other words, *figure it out in your head.*

Example: Find the LCD for 1/24, 1/12, 1/6, 1/3.

By looking at the numbers it is easy to see that 24 is divisible by 24, 12, 6 and 3. The LCD cannot be any smaller than the largest denominator, so 24 is the LCD.

Many LCDs are easily found by inspection. Look for even denominators and multiples of numbers as hints.

CONTINUED NEXT PAGE

Method 2.   **FACTOR TREE METHOD (same problem)**

Step 1.   Find the PRIME FACTORS of each of the denominators.

$$24 = 2 \times 2 \times 2 \times 3 = 2^3 \times 3^1$$

$$12 = 2 \times 2 \times 3 \quad = 2^2 \times 3^1$$

$$6 = 2 \times 3 \quad\quad = 2^1 \times 3^1$$

$$3 = 3 \quad\quad\quad = 3^1$$

Step 2.   Multiply together the *highest* power of each *different* base's factors.

The different bases are: 2 and 3
The highest power of 2 is 2^3, the highest power of 3 is 3^1

$$2^3 \times 3^1 = 24$$

If you use *all* the powers, the product will be too large.

SEE ALSO: COMMON DENOMINATOR   84

LEAST COMMON MULTIPLE

The *Least Common Multiple* (LCM) of a set of whole numbers is the *smallest* multiple that is common to the multiples of each number.

Example: What is the LCM of 5 and 6?

Multiples of 5 are: 5, 10, 15, 20, 25, 30 . . .
Multiples of 6 are: 6, 12, 18, 24, 30 . . .

The *smallest* multiple that is *common* to both sets is 30.

Finding the answer by inspection will not always be as easy as in the above example.   A method for determining the LCM follows:

Example: Find the LCM of 105 and 210.

Step 1.   Find the *prime factors* of each number and write in exponential form.

$105 = 3 \times 5 \times 7$ (all bases are first power)
$210 = 2 \times 3 \times 5 \times 7$ (all bases are first power)

Step 2. Multiply together the *highest* exponential of each different base.

$2 \times 3 \times 5 \times 7 = 210$

Example: Find the LCM of 24, 36 and 48.

Step 1. Find the prime factors of each number.

$24 = 2^3 \times 3$
$36 = 2^2 \times 3^2$
$48 = 2^4 \times 3$

Step 2. Multiply together only the *highest* exponent of each different base.

$2^4 \times 3^2 = 144$

If you use *every* exponential in the multiplication, the number will be too large.

SEE ALSO: GREATEST COMMON FACTOR

METRIC SYSTEM

The metric system is based on the *METER* which is 39.37 inches. The distance from the equator to either pole is 10 million meters. Liquid measure is based on the *LITER*, approximately 1.0567 quarts. Latin prefixes describe parts or multiples of a meter and liter in terms of POWERS OF TEN.

1000 meters is a KILOmeter
100 meters is a HECTOmeter
10 meters is a DECAmeter
1 meter is 39.37 inches
0.1 meter is a DECImeter
0.01 meter is a CENTImeter
0.001 meter is a MILLImeter

1000 liters is a KILOliter
100 liters is a HECTOliter
10 liters is a DECAliter
1 liter is 1.0567 quarts
0.1 liter is a DECIliter
0.01 liter is a CENTIliter
0.001 liter is a MILLIliter

It is a convenient fact that 1 milliliter (ml) = 1 cubic centimeter (cc) which relates liquid measure and volume in the metric system.

In the metric system weights are expressed in grams and kilograms. There are 1000 grams in one kilogram (often called a kilo.)

CONTINUED NEXT PAGE

Conversion ratios that are particularly handy are:

2.54 cm = 1 inch   **1.06 qts = 1 liter**   **2.2 lbs = 1 kg**

To convert one measure into another, multiply the given measure by a conversion ratio in which the undesired units divide out.

Example: Change 3 quarts into milliliters.

Step 1.   Start with what is given - 3 quarts.

Step 2.   Look for a conversion ratio that relates quarts and a metric measure.   1.06 quarts = 1 liter and 1 liter = 1000 ml

Step 3.   Set up the multiplication so that the units divide out.

$$3 \text{ quarts} \times \frac{1 \text{ liter}}{1.06 \text{ quarts}} \times \frac{1000 \text{ ml}}{1 \text{ liter}} = 2,830.2 \text{ ml}$$

Quarts in the numerator divide out with quarts in the denominator. Liters in the numerator divide out with liters in the denominator leaving milliliters as the desired unit.

MIXED NUMBERS - ADDITION

Mixed numbers have a *whole number* part and a *fraction* part.

Example: $4\frac{1}{2}$ is four PLUS one-half

It is NOT necessary to change mixed numbers into improper fractions in order to add them. First add the whole number parts and the fraction parts separately, then combine them.

Example: Add $5\frac{1}{3}$ and $2\frac{1}{5}$

Step 1.   Add the 5 and 2 to get 7 (whole number parts).

Step 2.   Add the 1/3 and 1/5 to get 8/15 (need a CD).

Step 3.   Put the parts together $7\frac{8}{15}$

If the fractional part of the answer turns out to be an improper fraction, change it to a mixed number and combine again.

CONTINUED NEXT PAGE   89

Example: Add $1 \frac{4}{5}$ and $6 \frac{2}{3}$

Step 1. Add the 1 and 6 to get 7 (whole number parts).

Step 2. Add 4/5 and 2/3 to get 22/15 (need a CD).

Step 3. Note that 22/15 is an improper fraction, so change it into a mixed number.

$$\frac{22}{15} = 1 \frac{7}{15}$$

Step 4. Add $1 \frac{7}{15} + 7$ to get $8 \frac{7}{15}$ (final answer)

SEE ALSO: COMMON DENOMINATOR,
IMPROPER FRACTIONS

MIXED NUMBERS - DIVISION

To divide mixed numbers it is necessary to have both numbers expressed as fractions.

Example: Divide $4\frac{1}{8} \div 2\frac{1}{5}$

Step 1. Change both mixed numbers into improper fractions.

$$\frac{33}{8} \div \frac{11}{5}$$

Step 2. Perform the division of fractions by multiplying the first fraction by the reciprocal of the second.

$$\frac{33}{8} \times \frac{5}{11} = \frac{15}{8}$$

Step 3. Simplify.

$$\frac{15}{8} = 1\frac{7}{8}$$

SEE ALSO: FRACTIONS - DIVISION, IMPROPER FRACTIONS   91

MIXED NUMBERS - MULTIPLICATION

To multiply mixed numbers it is necessary to have both numbers expressed as fractions.

Example: Multiply $4\frac{1}{2}$ X $3\frac{1}{3}$

Step 1.   Change both mixed numbers into improper fractions.

$$\frac{9}{2} \quad X \quad \frac{10}{3}$$

Step 2.   Multiply numerators; multiply denominators.

$$\frac{9}{2} \quad X \quad \frac{10}{3} \quad = \quad \frac{90}{6}$$

Step 3.   Simplify if desired.

$$\frac{90}{6} \quad = \quad 15$$

CONTINUED NEXT PAGE   93

Warning: You may NOT multiply the whole number parts together, then the fraction parts together to get the answer.

Example: If this were done for the previous problem, here is how the answer would work out.

$$4 \frac{1}{2} \times 3 \frac{1}{3} = 12 \frac{1}{6} \text{ which is INCORRECT}$$

SEE ALSO: FRACTIONS - MULTIPLICATION,
IMPROPER FRACTIONS

MIXED NUMBERS TO IMPROPER FRACTIONS

Definition: An *improper fraction* is one in which the numerator is greater than or equal to the denominator. There is nothing wrong with an improper fraction. These fractions behave the same way as proper fractions.

Example: Change $2\frac{1}{2}$ into an improper fraction.

Step 1. Multiply the whole number part by the denominator.

$2 \times 2 = 4$

Step 2. Add the numerator part.

$4 + 1 = 5$

Step 3. Write this number (5) over the denominator.

$$\frac{5}{2}$$

CONTINUED NEXT PAGE

This "magic formula" of multiplying the whole number part by the denominator of the fraction and then adding the numerator is exactly what is happening, but is difficult to see.

Example: Change $2\frac{1}{2}$ into an improper fraction.

Looking at the example, remember that a whole number may be expressed as a fraction.

Express the 2 as $\frac{2}{1}$ and change it into an equivalent fraction with the same denominator as the fraction part.

$$\frac{2}{1} = \frac{4}{2}$$

Add the two fractions together.

$$\frac{4}{2} + \frac{1}{2} = \frac{5}{2} \quad \text{which is the same answer.}$$

SEE ALSO: EQUIVALENT FRACTIONS

MIXED NUMBERS - SUBTRACTION

There are several ways to approach the subtraction of mixed numbers.
Method 1 is not the easiest to do, but is the easiest to remember.
Method 2 affords a deeper understanding of mixed numbers.

Example: Subtract $4\frac{1}{2} - 2\frac{5}{6}$

METHOD 1.

Step 1.   Change both mixed numbers into improper fractions.

$$\frac{9}{2} - \frac{17}{6}$$

Step 2.   Find a CD and write equivalent fractions.
Step 3.   Subtract numerators and simplify.

$$\frac{27}{6} - \frac{17}{6} = \frac{10}{6} = 1\frac{2}{3}$$

CONTINUED NEXT PAGE   97

METHOD 2.

Step 1.   Line up the subtraction vertically.
Step 2.   Find CD and write equivalent fractions.
Step 3.   Take 1 away from the 4, write it as 6/6 and
          add it to the 3/6 to make 9/6.

$$4 \frac{1}{2} \;=\; 4 \frac{3}{6} \;=\; 3 \frac{9}{6}$$

$$-\, 2 \frac{5}{6} \;=\; -\, 2 \frac{5}{6} \;=\; -\, 2 \frac{5}{6}$$

Step 4.   Do the subtraction. $1 \dfrac{4}{6} \;=\; 1 \dfrac{2}{3}$

When a mixed number is to be subtracted from a whole number,
change the whole number into a mixed number.

Example: Change 4 into a mixed number. $4 = 3 \dfrac{3}{3}$

SEE ALSO: CD, FRACTIONS - SUBTRACTION

MULTIPLICATION TABLE

1	2	3	4	5	6	7	8	9	10	11	12	13	14	15
2	4	6	8	10	12	14	16	18	20	22	24	26	28	30
3	6	9	12	15	18	21	24	27	30	33	36	39	42	45
4	8	12	16	20	24	28	32	36	40	44	48	52	56	60
5	10	15	20	25	30	35	40	45	50	55	60	65	70	75
6	12	18	24	30	36	42	48	54	60	66	72	78	84	90
7	14	21	28	35	42	49	56	63	70	77	84	91	98	105
8	16	24	32	40	48	56	64	72	80	88	96	104	112	120
9	18	27	36	45	54	63	72	81	90	99	108	117	126	135
10	20	30	40	50	60	70	80	90	100	110	120	130	140	150
11	22	33	44	55	66	77	88	99	110	121	132	143	154	165
12	24	36	48	60	72	84	96	108	120	132	144	156	168	180
13	26	39	52	65	78	91	104	117	130	143	156	169	182	195
14	28	42	56	70	84	98	112	126	140	154	168	182	196	210
15	30	45	60	75	90	105	120	135	150	165	180	195	210	225

Italics indicate the diagonal of perfect squares

CONTINUED NEXT PAGE

USES OF THE MULTIPLICATION TABLE

- Find products
- A division table to find quotients
- Find perfect squares
- Find square roots of perfect squares
- Find multiples of a number
- Find composite numbers
- Find patterns of numbers

It is not necessary to memorize the entire multiplication table. If you fold the table along the diagonal of perfect squares you will notice that one half is a mirror image of the other half.

You probably know most of the table already with the exception of a few "demons" like 6 x 9. Simply note the multiplication facts that you do not know and write these on 3 x 5 cards to use as flash cards to help you learn them.

SEE ALSO: COMPOSITE NUMBERS, PRIME NUMBERS,
            SQUARES

NEGATIVE NUMBERS

The number line extends infinitely in both directions, positive numbers to the right and negative numbers to the left.

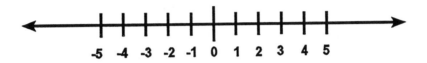

To ADD a negative number, move that many marks to the left. To ADD a positive number, move that many marks to the right.

Example: $(-5) + (-3) = -8$ (negative + negative is a negative)

$(8) + (-3) = 5$ (negative + positive has sign of larger)

To SUBTRACT negative numbers, DO NOT USE THE NUMBER LINE. Invoke the *definition* of subtraction which is to *ADD THE OPPOSITE.*

CONTINUED NEXT PAGE   101

Example: $3 - (-2) = 5$

Given three positive blocks (+)(+)(+) take away two negative blocks. There aren't any, so supply two zeroes in the form (+)(-) and (+)(-) to give you (+)(+)(+) (+)(-) (+)(-). Now you can take away two (-) to leave (+)(+)(+)(+)(+) or +5.

This concept is difficult to see arithmetically. The definition of subtraction is to *add the opposite,* so change the subtraction to an addition and make the (- 2) a positive 2.

$$3 + (+2) = +5$$

Multiplication and division involving negative numbers is very simple. If the signs of the numbers (not the operator signs) are *alike* the result is *positive.* If the signs of the numbers (not the operator signs) are *unlike*, the result is *negative.*

Examples: $(4)(5) = 20$ and $(-4)(-3) = 12$ (like signs)

$(-4)(5) = -20$ and $(6)(-3) = -18$ (unlike signs)

NUMBERS - KINDS OF

COUNTING NUMBERS start with 1, 2, 3, 4 .... These are the numbers that evolved in the accounting process driven by the demands of commerce.

DECIMAL NUMBERS are expressed as powers of ten such as 5.37 which is 5 and 37 hundredths. The decimal system is a *place* system where the position of the digits is significant. The fact that the 7 is two places to the right of the decimal point makes it 7/100.

FRACTIONS are numbers that express a division such as 3/4. The numerator (top) is divided by the denominator (bottom) to produce 0.75 a decimal quotient. The denominator of a fraction can never be zero.

INTEGERS are all the natural numbers, zero and the negatives of the natural numbers.

IRRATIONAL NUMBERS are those which cannot be expressed as fractions because they go on and on ... such as 5.34178 .... Pi is such a number. $\sqrt{5}$ is irrational because the decimal goes on and on.

CONTINUED NEXT PAGE

MIXED NUMBERS are composed of a whole number part and a fractional part. They may be separated into their respective parts.

NATURAL NUMBERS are the counting numbers. It is another name for the same set of numbers.

NEGATIVE NUMBERS are those numbers less than zero. They are found on the left hand side of the zero on the number line. They are expressed with a negative sign (-) such as - 6, - 9, - 3.5 and so on. Fractions may also be negative. There is a big difference between a minus sign and a subtraction sign. A minus sign indicates a negative number and a subtraction sign indicates the operation of subtraction.

RATIONAL NUMBERS are all those numbers that may be expressed as fractions. All whole numbers may be expressed as fractions by writing them over 1, therefore all whole numbers are rational. All terminal and repeating decimals are rational.

WHOLE NUMBERS are the set of counting numbers plus the zero which was added later -- 0, 1, 2, 3, 4, 5 . . . . Whole numbers are positive.

OPERATIONS AND OPERATORS

The *BASIC OPERATIONS* of arithmetic are adding, subtracting, multiplying and dividing. Other operations are squaring, cubing, taking a square root, a cube root and percent.

The *OPERATORS* of arithmetic are the symbols $+$, $-$, $\times$, $\div$, the radical sign ($\sqrt{}$), %, and exponents.

The *order* in which the operations are done is critical, and referred to as the ORDER OF OPERATIONS.

The operators *operate* on numbers. They permit numbers to interact with one another according to the rules of addition, subtraction, multiplication, division, etc.

Examples: In the problem $8 + 6 = 7$ the operation is addition of two positive integers.

In the problem $5 \times (6 - 2) = 20$ there are two operations. First comes the subtraction in the parentheses followed

CONTINUED NEXT PAGE

by multiplication of that result by 5.

When operating with negative numbers it is important to make a distinction between the sign of the number and the operator.

Example: In the problem [- 3][12 - (- 4)] there are several operations.

First: In the second set of [ ] there is the subtraction of a negative number.

$$12 - (- 4) = 16$$

Second: The negative 3 then multiplies the 16 to equal - 48

$$(- 3)(16) = - 48$$

SEE ALSO: NEGATIVE NUMBERS,
OPERATIONS - TERMINOLOGY,
ORDER OF OPERATIONS,
TERMS AND MEANINGS

OPERATIONS - TERMINOLOGY

OPERATION	Symbol	RESULT IS CALLED
Addition	$+$	Sum
Cubing	n^3	Cube
Division	$\div$	Quotient
Factoring	$(m)(n)$	Factors
Multiplication	X	Product
Square root	$\sqrt{}$	Square root
Squaring	n^2	Square
Subtraction	$-$	Difference

SEE ALSO: OPERATIONS AND OPERATORS,
SQUARE ROOT

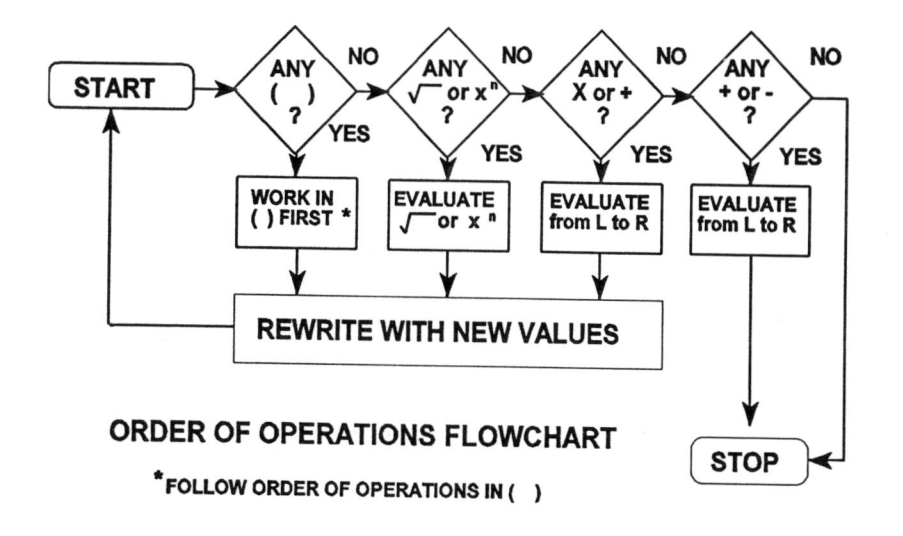

ORDER OF OPERATIONS FLOWCHART

*FOLLOW ORDER OF OPERATIONS IN ( )

108

ORDER OF OPERATIONS

The *order* in which operations are performed in arithmetic is CRUCIAL.
If the rules of order are not followed, calculations will come out wrong.
The *Order of Operations* is:

First: Do everything that is in parentheses following the order of operations.

Second: Evaluate numbers with exponents or radicals.

Third: Do all multiplications AND/OR divisions (whichever come first) from left to right.

Fourth: Do all additions AND/OR subtractions (whichever come first) from left to right.

Example: Evaluate $16 - 12 \div 2^2 \times 3 = 7$

Step 1. Check for ( ) - No parentheses, go to Step 2.

CONTINUED NEXT PAGE

Step 2.   Evaluate the exponential, 2 squared, which is 4.

$$16 - 12 \div 4 \times 3$$

Step 3.   Divide 12 by 4 to get 3.

$$16 - 3 \times 3$$

Step 4.   Multiply 3 by 3 to get 9.

$$16 - 9$$

Step 5.   Subtract 9 from 16 to get 7.

It is important to remember that you must do the multiplications
AND/OR divisions in the same step, NOT the multiplications first
followed by the divisions. The same caution applies to the additions
AND/OR subtractions. It is important also to go from left to right.

On a scale of importance of 1 to 10, Order of Operations is a BIG 10.

SEE ALSO: OPERATIONS AND OPERATORS   110

PERCENT

Symbol %

The percent sign (%) means to *divide by 100.* On most calculators the % button divides the number in the display by 100. The % sign can be treated as an *operator* -- anywhere you see the % sign you can divide the preceding number by 100.

Example: 40% means $40 \div 100 = 40/100 = 0.4$

40% of 50 means

$$\frac{40}{100} \times 50 = \frac{2000}{100} = 20 \quad \text{OR} \quad 0.4 \times 50 = 20$$

To "take the percent" of a number, just multiply the two numbers together and divide by 100.

Thinking of % as a division by 100 will help clarify a great deal of mystery about percents.

CONTINUED NEXT PAGE

BASE x RATE = PERCENTAGE is the fundamental equation of percent.

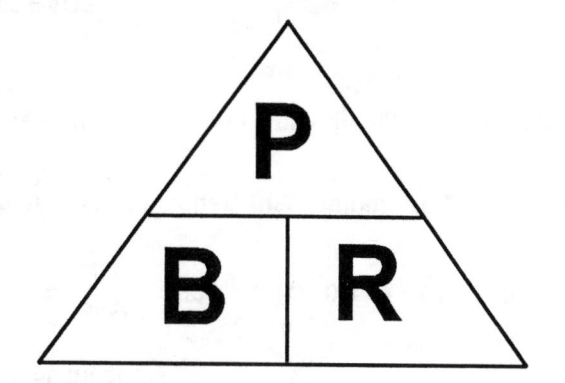

The % is *always* the RATE.

The PERCENTAGE is *always* a product -- the base x the rate.

The BASE is the *basis* of the problem such as the *principal* of a loan.

CONTINUED NEXT PAGE

Use the triangle above to find the quantity you are looking for.

If you want percentage, cover over the letter P and see that B and R are left uncovered. So, to find percentage, multiply BASE x RATE.

If you want rate, cover over the letter R and see that P and B are left uncovered. To find the rate, divide the PERCENTAGE by the BASE.

If you want the base, cover over the letter B and see that P and R are left. To find the base, divide the PERCENTAGE by the RATE.

Example: Base x Rate = Percentage

         30   x 10% =   3

If you were looking for the base, 3 divided by 10% = 30

If you were looking for the percentage, 30 x 0.1 = 3

If you were looking for the rate, 3 divided by 30 = 0.1 = 10%

CONTINUED NEXT PAGE

Example: What is 38% of 186?

The "what" is the number you are looking for.
The "is" stands for the = sign.
38 is the percent.
% means to divide by 100.
"of" means multiplication.

$$? = 38/100 \times 186$$

$$? = 0.38 \times 186$$

$$? = 70.68$$

Example: 10 is what % of 50?

$$10 = ?\% \times 50$$

$$\frac{10}{50} = ?\% = 0.20 = 20\%$$

PERCENT CHANGE

Definition: $$\frac{\text{The amount of change}}{\text{original value}} \times 100\% = \% \text{ CHANGE}$$

Example: The temperature changed from 33°F to 45°F in one hour. What was the percent change in temperature?

Step 1. Determine the amount of change (the difference between the two numbers).

45 - 33 = 12

Step 2. Determine the original amount.

The starting temperature (original) was 33°F.

Step 3. Divide the amount of change (12°F) by the original (33°F.)

12°F divided by 33°F = 0.364

CONTINUED NEXT PAGE

Step 4.   Multiply this decimal by 100% to get the percent change.

0.364 x 100% = 36.4% (rounded to nearest tenth)

Since the temperature went up, it is a percent *increase.*

Hints: In some problems the words *from* and *to* are used. The
          from part is the original.

          Whatever *comes first* in time is the original such as dates.

Example: Acme Manufacturing stock sold for $11.00 a share in
          November 1994 and today it is selling for $14.00 a share.
          What is the percent increase in stock price?

          The difference is $3 per share. The original price is $11.00.

          $$\frac{\$3.00}{\$11.00} \times 100\% = 27.3\%$$

SEE ALSO: DECIMALS TO PERCENT, PERCENT   116

PERCENT FRACTION

The *percent fraction* is simply a fraction that has a denominator of 100. By looking at the numerator you can determine the percent equivalent of the fraction.

Example: 60/100 = 60% because the % sign means to divide by 100. Whenever a fraction has a denominator of 100, the numerator is the percent.

Example: What is the % equivalent of 4/100?

4/100 = 4%

Example: What is the % equivalent of 5?

5 = 5/1 = 500/100 = 500%

Example: What is the % equivalent of 0.001?

0.001 = 1/1000 = 0.1/100 = one-tenth of one percent

SEE ALSO: FRACTIONS TO PERCENTS, PERCENT   117

PERCENTS TO DECIMALS

Rule: To change a percent to a decimal, divide the number by 100 and remove the % sign.

Example: Change 40% to a decimal.

40% = 40/100 = 0.4

What is happening here?

The % sign means to divide by 100. It is used as an operator. When the operation is done (dividing by 100) the % gets "used up," just like the + sign "disappears" when two numbers are added. In the problem 2 + 3 = 5, the plus sign gets "used up."

Example: Change 3% into a decimal.

3% = 3/100 = 0.03

The three forms, fraction, decimal and % are equivalents.

CONTINUED NEXT PAGE   119

SEE ALSO: **DECIMALS TO PERCENTS,**
EQUIVALENT FRACTIONS,
FRACTIONS TO PERCENTS,
PERCENT,
PERCENTS TO FRACTIONS

PLACE NAMES OF NUMBERS

9	2	5	7	3	8	.	4	1	6	7	3
HUNDREDS OF THOUSANDS	TENS OF THOUSANDS	THOUSANDS	HUNDREDS	TENS	UNITS		TENTHS	HUNDREDTHS	THOUSANDTHS	TEN-THOUSANDTHS	HUNDRED-THOUSANDTHS

Notes: -- Number places less than one end in *THS* such as ten*ths*.
-- The decimal point is read as "and."

CONTINUED NEXT PAGE

-- There is a hyphen in the ten-thousandths place to distinguish it from ten thousandths.

Example: 0.010 is ten thousandths while 0.0001 is one ten-thousandth.

-- There is an absence of symmetry in the decimal system. There is no "units" place on the right side of the decimal point to match the "units" on the left.

-- Each place in the decimal system is 10 times the value of the place to its right.

-- Each place in the decimal system is one-tenth of the value of the place to its left.

SEE ALSO: POWERS OF TEN   122

POWERS OF NUMBERS

Numbers may be "raised to powers." The little number to the upper right of a number is called its *POWER* (exponent).

Example: 2^3 is read as "Two raised to the third power."

Two is called the *BASE*. The 3 is the *POWER*. 2^3 means that the base 2 is used as a factor three times*, written out as:

$$2 \times 2 \times 2 = 8 \text{ }^*$$

So we read this as "two raised to the third power is eight."

There is a special name given to numbers raised to the second power. These numbers are referred to as *squares*. Numbers raised to the third power are called *cubes*.

* To say "two times itself three times" is incorrect. Two times itself is 4 (that's once), 4 x 2 = 8 (that's twice), 2 x 8 = 16 (that's three times).

CONTINUED NEXT PAGE   123

ANY NUMBER MAY BE RAISED TO ANY POWER

Whole numbers:	$4^3 = 64$; $0^3 = 0$
Decimals:	$1.3^2 = 1.69$; $0.02^3 = 0.000008$
Fractions:	$(1/2)^2 = 1/4$; $(2/3)^3 = 8/27$
Negative numbers:	$(-3)^2 = 9$; $(-2)^3 = -8$
Mixed numbers:	$(1\frac{1}{2})^2 = \frac{9}{4}$; $(2\frac{1}{3})^2 = \frac{49}{9}$
The ZERO power:	$4^0 = 1$; $n^0 = 1$
*Negative powers:	$2^{-3} = 1/8$; $3^{-2} = 1/9$
*Decimal powers:	$3^{1.2} = 3.74$; $4^{2.4} = 27.86$ (calculator)

*Advanced concept.

SEE ALSO: ZERO POWER OF A NUMBER

POWERS OF TEN

Symbol 10^n

Many practical problems involve multiplication or dividing by powers of ten. The decimal system, METRIC SYSTEM, ORDER OF MAGNITUDE, and currency calculations are examples.

Rule: When multiplying by *positive* powers of 10, move the decimal point ONE place to the right for EVERY factor of 10. The result will be larger number.

Examples: $60 \times 10^3 = 60 \times 10 \times 10 \times 10 = 60,000$

$0.5 \times 10^2 = 0.5 \times 10 \times 10 = 50$

The decimal was moved one place to the right for each factor of ten.

Rule: When dividing by *positive* powers of 10, move the decimal point ONE place to the left for EVERY factor of 10. The result will be a smaller number.

CONTINUED NEXT PAGE

Examples: $6,000 \div 10^3 = 6$

$138.2 \div 10^4 = 0.01382$

Q. Why do you "move decimals?"
A. To multiply or divide by powers of 10.

Q. When do you move decimals?
A. When dividing by a decimal; changing a decimal to a % or a % to a decimal.

Q. What does moving a decimal point do?
A. Moving it one place to the right makes the number 10 times larger. Moving it one place to the left makes the number 1/10 as large.

SEE ALSO: DECIMALS - DIVISION,
DECIMALS TO PERCENTS,
ORDER OF MAGNITUDE,
PERCENTS TO DECIMALS

PRIME NUMBERS - DEFINITION

Definition: A *prime number* is a whole number that can be evenly
              divided only by itself and 1.

Examples: 5 is prime
            7 is prime

            10 is *not* prime because it is *composed* of 2 x 5
            24 is *not* prime because it is a multiple of 2
            51 is *not* prime because it is composed of 3 x 17

The number 2 is the only *even* prime number. All other even numbers
are *composite* because they are multiples of 2.

A *prime* number cannot be expressed as the product of any other
numbers. Therefore, any number that appears in the multiplication
table can't be prime.

By definition, neither 1 nor 0 are prime numbers.

SEE ALSO: COMPOSITE NUMBERS, PRIME NUMBER TABLE   127

PRIME NUMBER TABLE

2	3	5	7	11	13	17	19	23	29	31	37
41	43	47	53	59	61	67	71	73	79	83	89
97	101	103	107	109	113	127	131	137	139	149	151
157	163	167	173	179	181	191	193	197	199	211	223
227	229	233	239	241	251	257	263	269	271	277	281
283	293	307	311	313	317	331	337	347	349	353	359
367	373	379	383	387	389	401	409	419	412	431	433
439	443	449	457	461	453	467	479	487	491	499	503
509	521	523	541	547	557	563	569	571	577	587	593
599	601	607	613	617	619	631	641	643	647	651	653
659	673	677	683	691	701	709	719	727	733	739	743
751	757	761	769	773	787	797	809	811	821	823	827
829	839	853	857	859	863	877	881	883	887	907	911
919	929	937	941	947	953	967	971	977	983	991	997

PROPORTIONS

Definition: A proportion is a comparison of two equivalent ratios.

A proportion is a set of two equivalent fractions.

A proportion has an = sign between two fractions.

Rule: The cross-products of a proportion are equal.

Example:
$$\frac{1}{4} = \frac{6}{24}$$

The two cross-products are $1 \times 24 = 24$ and $4 \times 6 = 24$ which, by inspection, are equal.

Most proportion problems involve finding a missing number.

Example:
$$\frac{?}{12} = \frac{8}{16}$$

CONTINUED NEXT PAGE

Step 1.   Multiply 12 x 8 = 96

Step 2.   Divide the 96 by 16 to get 6.

Now, this problem is easily done by inspection. If 8/16 = 1/2 then the first fraction must also be equal to 1/2. So, what must the numerator of the first fraction be? Obviously it is 6.

Another idea that comes in handy is the fact that any two factors may be COMMUTED, i.e., their places exchanged.

Example: To find the missing value, exchange the places of the X and the 66.

$$\frac{22}{X} = \frac{66}{99} \qquad = \qquad \frac{22}{66} = \frac{X}{99}$$

Now proceed as before. 22 x 99 = 2178; divide 2178 by 66 to get 33.

When proportions include units, it is important to include the units in the calculations.

SEE ALSO: RATIOS

PYTHAGOREAN THEOREM

Rule: *The sum of the squares of the sides of a right triangle is equal to the square of its hypotenuse,* expressed by the formula:

$$a^2 + b^2 = c^2$$

where a and b are sides and c is the hypotenuse.

Uses: If two parts of a triangle are known, the third can be calculated. Used in navigation, surveying, engineering, construction.

Example: One side of a triangle is 3 m (a) the other side is 4 m (b). What is the length of the hypotenuse (c)?

$$3^2 + 4^2 = c^2$$

$$9 + 16 = 25$$

$$c = 5$$

CONTINUED NEXT PAGE   131

Example: The hypotenuse of a right triangle is 13 feet. One of the legs (sides) of the triangle is 12 feet. What is the length of the other side?

$$a^2 + b^2 = c^2$$

Step 1. Substitute values.

$$(12\text{ ft})^2 + (\ ?\ )^2 = (13\text{ ft})^2$$

Step 2. Square each value.

$$144\text{ ft}^2 + (\ ?\ )^2 = 169\text{ ft}^2$$

Step 3. Subtract 144 ft² from each side of the equation.

$$(\ ?\ )^2 = 25\text{ ft}^2$$

Step 4. Take square root of both sides.

$$?\ =\ 5\text{ ft}\ =\ b$$

SEE ALSO: SQUARES

132

RADICALS

Symbol: $\sqrt[n]{}$

The radical sign $\left(\sqrt{} \right)$ represents the operation of extracting the *root* of a number.

Although there are many roots of a number, the common roots are square and cube roots.

The radical sign is composed of two parts, the bracket itself $\sqrt{}$ and an *index* which is a little number (n) found in the "v" of the radical sign. If there is no index shown, it is assumed that the index is 2 which means you are looking for the second root of the number, commonly called the *square root*.

Example: $\sqrt{36} = 6$ "the square root of 36 is 6"

Example: $\sqrt[3]{8} = 2$ "the cube root of 8 is 2"

The square root of a number is the reverse of the square of a number. The cube root of a number is the reverse of the cube of a number.

CONTINUED NEXT PAGE

RATIO

Definition: A ratio is a division expressed as a fraction. The fraction may be proper, improper or complex. The rules of fractions apply to ratios.

RATIOS are COMPARISONS.

Example: There are 4 men and 5 women in a group.

The ratio of MEN to WOMEN is: $\dfrac{4 \text{ men}}{5 \text{ women}}$

The ratio of WOMEN to MEN is: $\dfrac{5 \text{ women}}{4 \text{ men}}$

The ratio of MEN to TOTAL is: $\dfrac{4 \text{ men}}{9 \text{ people}}$

It is important to include the units in the ratios and to use the units in the calculations.

CONTINUED NEXT PAGE

Ratios may be made up of COMPLEX FRACTIONS such as 1/8 ounce to 4/5 pound.

$$\frac{\frac{1}{8} \text{ ounce}}{\frac{4}{5} \text{ pound}}$$

To simplify such a ratio:

Step 1. Work on the COMPLEX FRACTION first by doing the indicated division.

$$\frac{\frac{1}{8}}{\frac{4}{5}} = \frac{1}{8} \times \frac{5}{4} = \frac{5}{32}$$

Step 2. Write the new fraction with the old units.

$$\frac{5 \text{ ounces}}{32 \text{ pounds}}$$

SEE ALSO: COMPLEX FRACTIONS, CONVERSION RATIOS 136

RECIPROCALS

It is easier to demonstrate reciprocals than to define them.

$\dfrac{2}{3}$ is the reciprocal of $\dfrac{3}{2}$

$\dfrac{3}{2}$ is the reciprocal of $\dfrac{2}{3}$

2 is the reciprocal of $\dfrac{1}{2}$ (because 2 = 2/1)

Reciprocals occur in *pairs*. One number is the *reciprocal* of the other.

Definition: A number multiplied by its reciprocal = 1

$$\frac{2}{3} \times \frac{3}{2} = 1$$

The reciprocal concept is used frequently in mathematics.

CONTINUED NEXT PAGE   137

USES OF RECIPROCALS:

To divide one fraction by another, multiply the first fraction by the reciprocal of the second.

To clear a fraction from an equation, multiply through by the reciprocal of the fraction.

To change any fraction into 1, multiply it by its reciprocal.

The reciprocal concept is one of those ideas that gets more valuable the more you think about it. For example, the statement:

Multiplying by 1/2 is the same as dividing by 2

is full of different mathematical ideas. It relates division and multiplication of fractions. It contains the notion of reciprocals. It is a mathematical truth that needs a great deal of thought to master.

SEE ALSO: FRACTIONS - DIVISION

ROUNDING DECIMALS

Task: Round 4.106 to the nearest "*tenth*."

Skills: You must know PLACE NAMES OF NUMBERS.

Step 1. Locate the place to be rounded.

4.106

Step 2. Look to the neighbor to the right (0).

Step 3. If that neighbor is 5 or greater, raise the value of that digit one number higher. In this case 0 is less than 5.

Step 4. If that neighbor is less than 5, drop all the digits following the place to be rounded.

Answer: 4.1

CONTINUED NEXT PAGE

Example: Round 4.875 to the nearest *tenth.*

The *tenth* place is the digit 8.

The neighbor to the right of 8 is a 7.

7 is larger than 5.

Round *UP* to the higher digit.   Ans: 4.9

Example: Round 4.875 to the nearest *hundredth.*

The *hundredth* place is a 7.

The neighbor to the right is a 5.

Round *UP* to the higher digit.   Ans: 4.88

SEE ALSO: ROUNDING WHOLE NUMBERS   140

ROUNDING WHOLE NUMBERS

Task: Round 2546 to the nearest *hundred.*

Skills: You must know the PLACE NAMES OF NUMBERS and how
to put numbers in order.

Step 1.   Find the nearest hundred *smaller* than 3546 and the nearest
hundred *larger* than 2546.

> 2500   nearest *smaller* hundred
> *2546*
> 2600   nearest *larger* hundred

Now, by looking at the numbers, determine whether 2546 is closer to
2500 or to 2600. The number being tested is closer to 2500, therefore
we say, "2546 rounded to the nearest hundred is 2500."

If the number falls right in the *middle*, such as 2550, the rule states:
round to the higher value.

CONTINUED NEXT PAGE

ROUNDING RULE OF THUMB

Rule: To round a whole number, find the place to be rounded and look at the digit to its right. If that digit is 5 or greater, round up (to the higher number). If that digit is less than 5, change that digit and the remaining digits to the right to zeroes.

Example: Round 2546 to the nearest *hundred.*

Step 1. Go to the hundreds place. It is the 5.

Step 2. Inspect the digit to its right. It is the 4.

Step 3. Since 4 is less than 5, round down to the lower number.

Step 4. Change the 4 and the remaining digits to the right to zeroes.

2546 rounded to the nearest hundred is 2500.

SEE ALSO: PLACE NAMES OF NUMBERS   142

SCIENTIFIC NOTATION

Scientific Notation is a method of expressing numbers in terms of powers of ten. This is particularly useful when dealing with very large or very small numbers. There are 602,000,000,000,000,000,000,000 molecules of H_2O in 18 grams of water (Avogadro's Number.) A more convenient way to write the number is 6.02×10^{23}.

Example: Write 567,334,563,060 in scientific notation.

Step 1. Locate the decimal point in the original number. In this case it is all the way to the right of the last zero.

$$567,334,563,060.$$

Step 2. Place a *new* decimal point to the right of the first digit on the left end of the number. In this case it is between the 5 and the 6.

$$5.67\ 334\ 563\ 060.$$

Step 3. Count the number of places you had to move

from right to left. In this case it is 11 places.

$$5.\underline{67\ 334\ 563\ 060}.$$
11 places

Step 4.   This number is the exponent on the 10, i.e., 10^{11}

Step 5.   Round off the remainder of the number to three significant figures, and multiply that number by the power of 10.

$$5.67 \times 10^{11}$$

If you are dealing with a number less than one, you will be moving the decimal point in the opposite direction and the exponent will be negative.

Example: Write 0.00004527 in scientific notation.

$$4.53 \times 10^{-5}$$

SEE ALSO: ORDER OF MAGNITUDE,
            PLACE NAMES OF NUMBERS   144

SIMPLIFYING (fractions, mixed numbers, etc.)

To "simplify" in mathematics means many things.

FRACTIONS: *Simplify* means to express the fraction in lowest terms, i.e., with a numerator and denominator that is not divisible by the same whole number.

Example: Simplify $\frac{106}{240}$ to lowest terms.

Step 1. Use the FACTOR TREE method to find the prime factors of the numerator and the denominator.

$$\frac{106}{204} = \frac{2 \times 53}{2 \times 2 \times 3 \times 17}$$

Step 2. Divide out the common factors from the numerator and denominator and then multiply all the remaining factors.

$$\frac{2 \times 53}{2 \times 2 \times 3 \times 17} = \frac{53}{102}$$

CONTINUED NEXT PAGE   145

IMPROPER FRACTIONS: Change the improper fraction into a mixed number by dividing the numerator by the denominator and expressing the remainder as a fraction.

Example: $\dfrac{22}{5} = 4\dfrac{2}{5}$

NUMERICAL EXPRESSIONS: Perform as many of the indicated operations as possible following the order of operations. When no more operations can be performed, the expression is said to be evaluated, i.e., expressed in simplest numerical form.

Example: Evaluate the expression (5 x 6 - 3 + 8)

Using the order of operations, the expression equals 35.

SEE ALSO: FRACTIONS - DIVISION,
IMPROPER FRACTIONS,
MIXED NUMBER TO IMPROPER FRACTIONS,
ORDER OF OPERATIONS

SQUARES

A square may be a geometrical figure that has all four sides equal and all the angles are right (90°) angles. A square may also be the result of multiplying a number by itself. There is an important connection between these two definitions.

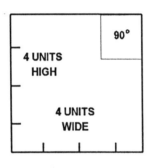

In the equation **4 x 4 = 16**, 16 is said to be the "square" of **4**.

The geometric square to the left has a side of 4 units and an area of 16 square units.

The area of a square can be written S^2.

The multiplication table has a *diagonal of perfect squares* that runs from the upper left to the lower right of the table.

CONTINUED NEXT PAGE   147

■ A number is squared to find the area of the square when given one side.

■ Many physical laws in nature obey the *square rule* such as the way the intensity of light varies with distance.

■ Distances are often calculated by squaring the sides of a right triangle using the PYTHAGOREAN THEOREM.

■ In construction, squaring is used to make sure that angles in buildings are "square" such as foundations, walls, and pilings.

SEE ALSO: MULTIPLICATION TABLE,
               SQUARE AND RECTANGULAR NUMBERS,
               SQUARE ROOTS   148

SQUARE AND RECTANGULAR NUMBERS

If you express a number in terms of the geometric figure that it defines, you can tell if a number is prime, rectangular or square.

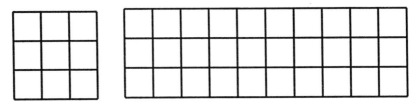

9 is a square number   30 is a rectangular number

17 is a prime number

CONTINUED NEXT PAGE

A prime number can only be arranged as a rectangle with one of the sides equal to 1. It cannot be arranged as a square or as a rectangle with both sides greater than 1. If a number can be arranged as a rectangle or a square it cannot be prime.

A rectangular number cannot be arranged as a square, although a square number may be arranged as a rectangle.

The *AREA* of a rectangle is *length times width.*

Decimals can also be squared. 3.1 x 3.1 = 9.61
But be careful with the decimal place.

The squares of numbers can be found in SQUARE ROOTS - Table of Squares and Roots.

SEE ALSO: SQUARES, SQUARE ROOTS - Definitions,
            SQUARE ROOTS - How To Find Them

SQUARE ROOTS - Definitions   Symbol: $\sqrt{}$

The *square root* of a number may be thought of as the side of a square that has an area equal to the number.

Example: A square with an area
            of 25 square units has
            a side of 5 units.

The square root of a number may also be thought of as the reverse of squaring. The number 4 squared is 16. The square root of 16 is 4.

Another way to think about a square root is to answer this question: What number multiplied by itself is 49?   Ans.   7. Therefore 7 is the square root of 49.

25 sq units

side = 5

CONTINUED NEXT PAGE   151

The "common square roots" are the roots of the "perfect squares." These roots also happen to be the counting numbers.

$$\sqrt{1} = 1 \qquad \sqrt{16} = 4$$

$$\sqrt{4} = 2 \qquad \sqrt{25} = 5$$

$$\sqrt{9} = 3 \qquad \sqrt{36} = 6$$

Note: A square root is not a division. Look closely at the difference between a square root sign and the division symbol.

$$\sqrt{} \qquad\qquad\qquad \overline{)} \ \text{ or } \ \overline{\lceil}$$

Square root   Division

SEE ALSO: Next four pages

SQUARE ROOTS - Helpful Hints

Many square roots do not work out evenly. $\sqrt{5}$ = 2.236... It keeps going and going and going. How far you carry it out depends on the precision that you need for your calculations. For normal use, round off to two decimal places. $\sqrt{5}$ = 2.24

Some square roots look difficult, but are really simple squares in disguise.

Example: What is the square root of 1.44?

Notice that without the decimal point the number would be 144, a perfect square. So try 1.2 x 1.2. It works out. However, the square root of 14.4 does not come out evenly. Try it on a calculator.

The *diagonal of perfect squares* in the multiplication table is also a square root table of the perfect squares. Locate a number on the diagonal and its square root will be found at the top of the column and at the left hand side of the row.

CONTINUED NEXT PAGE

SQUARE ROOTS - How to find them

A square root may be found in a number of ways:

- Look it up in a *Table of Square Roots.*

- Use a hand calculator. Enter the number and depress the $\sqrt{}$ key. (Do not use the x^2 key - this is the *squaring* key.)

- Raise the number to the 0.5 power on the calculator (you will need the y^x key.) Enter the number, depress the y^x key, enter 0.5 and then depress the = key.

- Approximate by guessing, then square the guess to see how close it comes to the number. Make a better approximation and repeat the process.

- For fractions - find the square root of the numerator and the square root of the denominator.

SEE ALSO: SQUARE ROOTS - Table of Squares and Roots   154

SQUARE ROOTS - Table of Squares and Roots

No	Square	Square Root		No	Square	Square Root
1	1	1		10	100	3.16
2	4	1.41		11	121	3.32
3	9	1.73		12	144	3.46
4	16	2		13	169	3.61
5	25	2.24		14	196	3.74
6	36	2.45		15	225	3.87
7	49	2.65		16	256	4
8	64	2.83		17	289	4.12
9	81	3		18	324	4.24

CONTINUED NEXT PAGE

No	Square	Square Root		No	Square	Square Root
19	361	4.36		28	784	5.29
20	400	4.47		29	841	5.39
21	441	4.58		30	900	5.48
22	484	4.69		31	961	5.57
23	529	4.80		32	1024	5.66
24	576	4.90		33	1089	5.74
25	625	5		34	1156	5.83
26	676	5.10		35	1225	5.92
27	729	5.20		36	1296	6

You may use a calculator to find the square or the square root of a number.

SURFACE AREA

The *surface area* of any solid figure is the *sum* of all the different areas on the outside of the volume. There are formulas for different figures, but almost any surface area can be calculated by adding up the smaller parts.

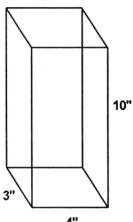

Example: Find the surface area of a rect-angular solid that is 10 inches x 3 inches x 4 inches.

Step 1.   Make a sketch of the solid figure and write in the dimensions.

Step 3.   Notice that there are six sides. These sides come in pairs: front/back, side/side, top/bottom.

Calculate the area of the top.
Calculate the area of a side.
Calculate the area of the front.

CONTINUED NEXT PAGE   157

Area of the top is	12 square inches
Area of side is	30 square inches
Area of front is	40 square inches
Sum is	82 square inches

Step 4.   Multiply this number by two because there are two of each face.

82 square inches x 2 = 164 square inches.

Some surface area formulas:

Cube	$6s^2$	6 times the square of an edge
Sphere	$4\pi r^2$	4 times π times the square of the radius

It also helps to visualize the solid as a box and unfold it, flattening it out and then calculating the smaller areas.

SEE ALSO: AREA

TERMS AND MEANINGS

Many English words used in mathematics have strict technical meanings. It is important and helpful to understand these terms.

WORD	MATHEMATICAL MEANING	EXAMPLE
and	addition	4 *and* 6 is 10
consecutive	one after another	2, 3, 4, 5 . . .
decreased	subtraction	8 *decreased* by 3 is 5
difference	result of subtracting	*difference* of 7 and 2 is 5
increased	addition	3 *increased* by 4 is 7
is	equals	2 x 3 *is* 6
less	subtraction	9 *less* 6 is 3

CONTINUED NEXT PAGE

159

WORD	MATHEMATICAL MEANING	EXAMPLE
of	multiplication	3% *of* 60 is 1.8
per	division	12 gal *per* mile
product	result of multiplying	the *product* of 4 and 7 is 28
quotient	result of dividing	the *quotient* of 6 and 3 is 2
square	multiplication	3 *squared* is 9
sum	result of adding	the *sum* of 2 and 3 is 5
times	multiplication	8 *times* 9 is 72

TRIANGLES

Triangles are three-sided closed figures. Although triangles may be oriented in all directions, one side is usually referred to as the *base* and is drawn horizontally. The other sides are expressed in relation to the base.

The *height* of the triangle is a line that is dropped from the highest point (a vertex) and is perpendicular to the base. This line is also called the *altitude*.

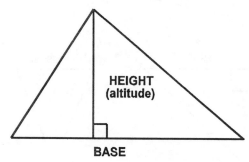

HEIGHT (altitude)

BASE

Triangles have different names depending on their properties. A triangle that has *three equal sides* is called *equilateral*. A triangle that

CONTINUED NEXT PAGE

has *two equal sides* is called *isosceles.* A triangle that contains one *right angle* (90°) is called a *right* triangle. Any triangle in which no two sides are equal is called *scalene.* Any triangle with an angle larger than 90° is called *obtuse.*

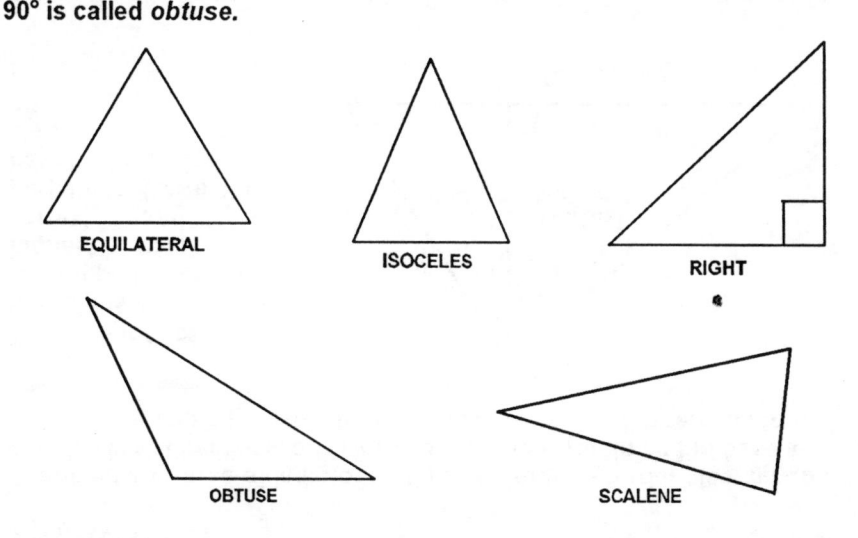

EQUILATERAL

ISOCELES

RIGHT

OBTUSE

SCALENE

VOLUME

Symbol V

Volume has three dimensions, length, width and height. It is a measure of the amount of space taken up by an object. Volumes are expressed in cubic units such as cubic feet (ft^3,) cubic centimeters (cm^3 or cc,) quarts, gallons, barrels. The units are *cubed.* Volume is different from *area.* Area has two dimensions. Volume has three dimensions.

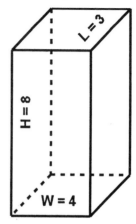

There are many formulas that can be used to calculate the volume of a figure, but a good general idea is that:

AREA of the Base x Height = Volume

It does not matter what the shape of the base happens to be. In the figure to the right, the area of the base is 3 x 4 = 12 sq units. The height is 8 units. 12 sq units x 8 units = 96 cubic units. (in x in x in = in^3)

CONTINUED NEXT PAGE

If the area of the base happens to be a triangle, calculate the area of the triangle and multiply by the height.

If the solid happens to be a cylinder, calculate the area of the circle of the base and multiply by the height.

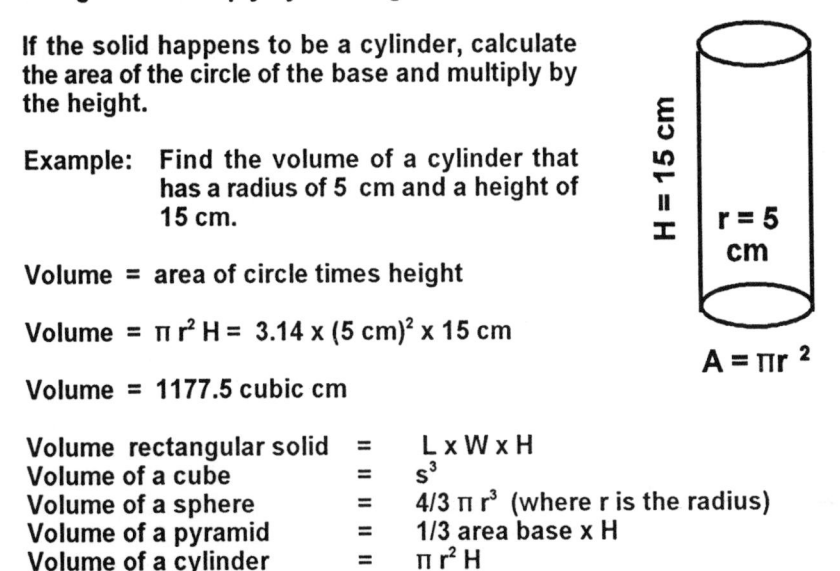

Example: Find the volume of a cylinder that has a radius of 5 cm and a height of 15 cm.

Volume = area of circle times height

Volume = $\pi r^2 H$ = 3.14 x (5 cm)^2 x 15 cm

Volume = 1177.5 cubic cm

Volume rectangular solid	=	L x W x H
Volume of a cube	=	s^3
Volume of a sphere	=	4/3 πr^3 (where r is the radius)
Volume of a pyramid	=	1/3 area base x H
Volume of a cylinder	=	$\pi r^2 H$

164

ZERO FACTS

Zero is neither negative or positive, but it is an even number because it is located between two odd numbers, - 1 and + 1.

ZERO divided by any other number is ZERO.

Example: 0/3 = 0

ZERO multiplied by any other number is ZERO.

Example: 0 x 5 = 0

Division by ZERO is not defined. Do NOT divide by ZERO.

CONTINUED NEXT PAGE   165

Any number raised to the ZERO power is 1.

Example: $5^0 = 1$ (SEE ZERO POWER OF A NUMBER)

ZERO added to any other number leaves that number unchanged.

Example: $6 + 0 = 6$

ZEROES are used as "place-keepers" when the decimal point is moved beyond existing digits in a number.

Example: $0.16 \div 100 = 0.0016$

A leading ZERO is used in a decimal to alert the reader to look for a decimal point: 0.143 rather than .143 (both are correct but this form is preferable in math used for science and engineering).

SEE ALSO: ZERO POWER OF A NUMBER